THE LABCOAT GUIDE TO CROP PROTECTION R&D

RESEARCH & DEVELOPMENT COMPENDIUM

HARALD B. TEICHER

BIOCOMM

"I think the biggest innovations of the 21st century will be at the intersection of biology and technology. A new era is beginning."

- STEVE JOBS (WALTER ISAACSON)

ABOUT THIS BOOK

THE LABCOAT GUIDE TO CROP PROTECTION R&D is the compiled version of the first three books in the LABCOAT GUIDE TO CROP PROTECTION series.

Aimed at Crop Protection students, professionals, and others wishing to understand basic aspects of Crop Protection Development, this book is an easily accessible introduction to essential principles of Pesticide and Biopesticide Mode Of Action and Formulation, Biostatistics and Strategic R&D Management for Pesticide & BioPesticide R&D.

ABOUT THE AUTHOR

Harry Teicher is a Crop Protection Scientist and Plant Biochemist, providing independent Strategic R&D Management, Scientific & Communications solutions to AgChem and BioScience Organizations developing science-based products.

He has more than 20 years of Conventional & Biological AgChem R&D experience, with more than 10 years as Study Director providing biological & registration support for commercial crop protection products.

Harry Teicher is an Administrator of the INSEAD Strategic R&D Management Alumni Network LinkedIn group. Follow him on LinkedIn, Twitter and Facebook.

For more information:
www.biocomm.eu
teicher@biocomm.eu

BOOKS BY HARRY TEICHER

Starting a new Crop Protection project or starting a new role – the LABCOAT GUIDE TO CROP PROTECTION series will help you focus on specific topics and get you up to speed fast!

Available in eBook and Print formats at:

https://biocomm.eu/crop-protection-books/

PESTICIDES & BIOPESTICIDES: FORMULATION & MODE OF ACTION is the first book in the LABCOAT GUIDE TO CROP PROTECTION series - this book is an easily accessible introduction to essential principles of Pesticide and Biopesticide Mode Of Action and Formulation.

STRATEGIC R&D MANAGEMENT: AGCHEM & BIOSCIENCE is the second book in the LABCOAT GUIDE TO CROP PROTECTION series - this book is an easily accessible introduction to essential principles of Strategic R&D Management.

ESSENTIAL BIOSTATISTICS is the third book in the LABCOAT GUIDE TO CROP PROTECTION series - this book is an easily accessible primer for scientists and research workers not trained in mathematical theory, but who have previously followed a course in Biological Statistics.

ONLINE COURSES

Are you starting a new Crop Protection project, moving teams, or starting a new role - BIOSCIENCE SOLUTION'S ACADEMY course series will provide an easily accessible introduction to specific topics and get you up to speed fast!

These courses are relevant to anyone involved in Crop Protection R&D, including Corporate and Academic Researchers, R&D Managers, Registration Specialists, Product Managers, Business Analysts and Investors.

Available online at:

https://bioscience_academy.teachable.com/

ABOUT BIOSCIENCE SOLUTIONS

BIOCOMM.EU

SCIENTIFIC & STRATEGIC R&D MANAGEMENT
SCIENCE COMMUNICATIONS

Started in 2002 by Harry Teicher, BioScience Solutions provides Strategic R&D Management and Support as well as Communications solutions to organizations developing science-based products.

Visit our website at: biocomm.eu

CONTENTS

PESTICIDES & BIOPESTICIDES

ESSENTIAL BIOSTATISTICS

PART I
PROPOSING A HYPOTHESIS

PART II
DESIGNING AND IMPLEMENTING EXPERIMENTS (SIGNIFICANCE, POWER AND EFFECT)

PART III
DESCRIPTIVE STATISTICS - CRITICALLY EVALUATING EXPERIMENTAL DATA

PART IV
INFERENTIAL STATISTICS - ACCEPTING OR REJECTING THE HYPOTHESIS

PESTICIDES & BIOPESTICIDES

MODE OF ACTION & FORMULATION

INTRODUCTION TO PESTICIDES & BIOPESTICIDES

Crop protection refers to the measures taken to protect crops from pests, diseases, and other threats that can damage or destroy the crop. While crop protection is essential for ensuring the productivity and sustainability of agriculture, it is also facing a number of challenges.

One major challenge is the increasing resistance of pests and diseases to chemical pesticides. As pests and diseases become resistant to these chemicals, it becomes more difficult to control them, leading to reduced crop yields and increased production costs. To address this challenge, farmers and researchers are exploring alternative methods of crop protection, such as integrated pest management (IPM) and the use of biopesticides.

Another challenge is the impact of climate change on agriculture. Rising temperatures and changing weather patterns can lead to more frequent and severe pest and disease outbreaks, as well as reduce crop yields. This can make it more difficult to protect crops and can lead to increased food insecurity.

A third challenge is the limited availability of land for agriculture. As the global population continues to grow, there is increasing pressure on land resources, which can make it difficult for farmers to expand their operations or to protect their crops from external threats.

To address these challenges, there is a need for innovative solutions and approaches to crop protection.

This may include the development of new technologies and methods for controlling pests and diseases, the promotion of sustainable farming practices, and the creation of policies and regulations that support the protection of crops. By addressing these challenges, we can ensure the productivity and sustainability of agriculture in the face of changing conditions.

Global human population growth is around 75 million annually, and estimates place our total population at over 9 billion by mid-2050. The logistical challenge is not only the projected 9 billion people, but also the estimated hundred billion head of livestock we will need to feed to maintain our current consumption standards.

It has been estimated that there are only about 4 billion hectares of arable land in the world.

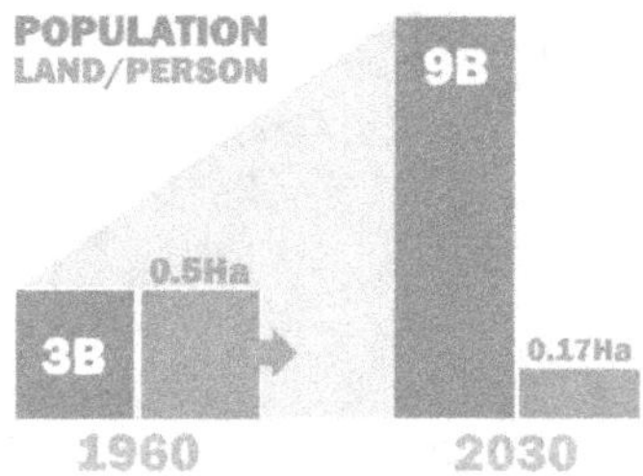

Figure I: Global population growth (1960 - 2030) vs. tillable land availability (Ha / person).

If 0.1 Ha of land is required to provide food for one person per year, then 4 billion Ha of tillable land would support 40 billion people. With a 1% annual population growth rate, the carrying capacity of Earth may be reached in a little over two centuries.

Without crop protection and plant enhancement strategies, we risk losing almost half of our overall crop production to insects, weeds and disease:

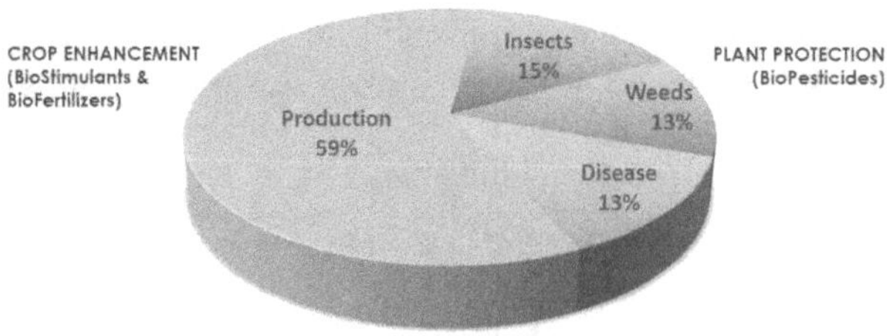

Figure II: Percentage of overall crop production at risk to loss by insects, weeds and disease.

Pesticides are agents, substances or mixtures of substances used to prevent, destroy or control pests (including weeds) and diseases.

A pesticide may be a chemical (*conventional pesticide*) or a biological agent (*biopesticide*).

Biopesticides, together with biostimulants and biofertilizers make up the product group of *Agricultural Biologicals*:

Biostimulants and biofertilizers are defined as biological substances or microorganisms capable of promoting crop health by increasing the availability of nutrients, or by acting as plant strengtheners and phytostimulants.

Biorational products are typically derived from natural or biological origins and include biological pesticides as well as products used for crop stress management, enhanced plant physiology benefits, and root growth management.

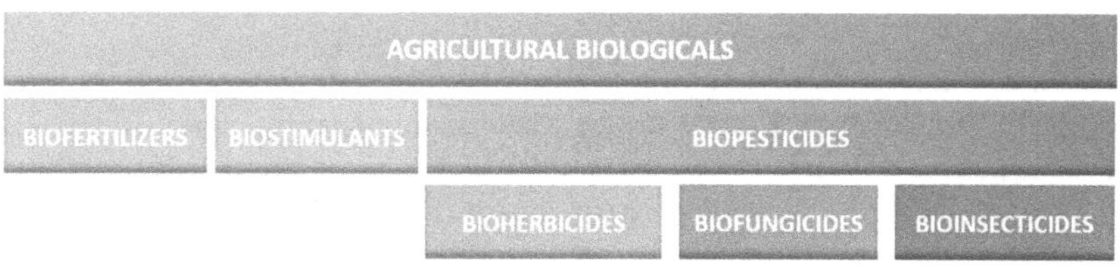

Figure III: Agricultural Biologicals.

Biopesticides introduce unique and complex modes of action, and can be used in plant production on an equal footing with chemical pesticides - if used correctly. Most biopesticides are intended for use with traditional pesticides, which is contributing to the rapid increase in their use.

Estimates of global market sizes for crop protection products (conventional pesticides as well as biopesticides) vary, but the consensus is that biopesticides are expected to increase from about 2% of the global pesticide market in 2003 to about 8% of the global pesticide market in 2020.

Current opinion is that the compound annual growth rate (CAGR) for biopesticides during this period is expected to exceed 12%, compared to an expected overall pesticide CAGR of about 7%.

Within the EU, stricter pesticide regulatory requirements, consumer awareness regarding hazards caused by chemical pesticides and the phasing out of toxic conventional pesticides are factors driving the growth of the biopesticide market.

Biopesticides are generally biodegradable, leave few (or no) harmful residues (easier residue management) and are less likely to harm non-target species.

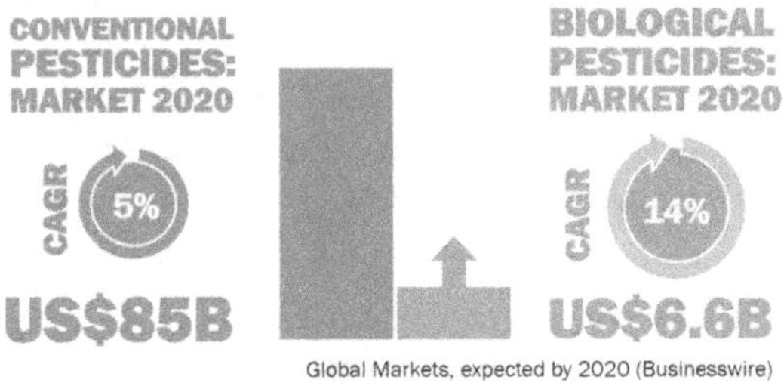

Figure IV: Market share and growth rates for Pesticides and BioPesticides.

However, biopesticides are not necessarily harmless, despite being of natural origin - the formation of plant-produced toxins induced by biopesticides is a case in point, while pyrethrins, spinosad and abamectin are registered by the EPA as conventional (chemical) pesticides due to their neurotoxic mode of action.

Biopesticides fall into three major classes:

Microbial pesticides, including the biopesticidal metabolites produced by these organisms

Bio-derived chemicals (metabolites) - naturally occurring compounds, including: *Semiochemicals* - natural compounds such as pheromones, that affect the behavior of other individuals; *Biocontrol compounds isolated from microorganisms; Biocontrol compounds isolated from fungi or crustacean shells,* such as Chitosan; *Plant extracts,* including secondary biocontrol metabolites such as alkaloids, terpenoids and phenolics; and *naturally occurring minerals* with biocontrol activity

Plant-incorporated protectants (PIPs) comprising foreign DNA inserted into crop genetic material (GM crops)

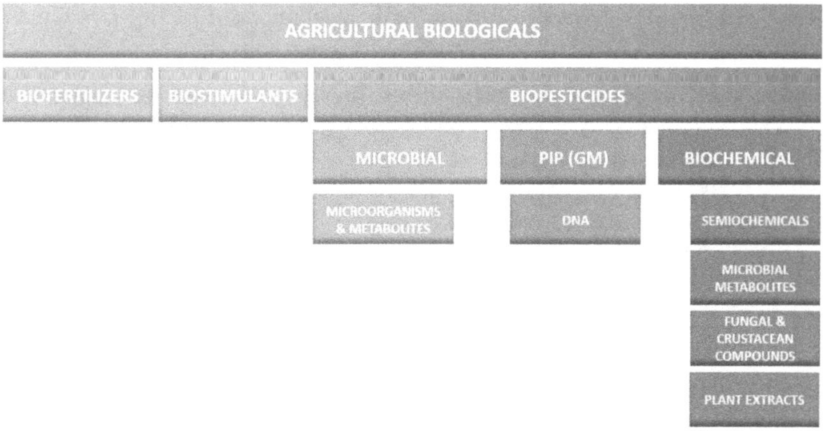

By introducing unique modes of action, biopesticides can be used in integrated strategies with traditional pesticides to increase crop yields by working synergistically with chemical pesticides, to extend application timings and allow timely reuptake intervals as well as to provide resistance strategies.

Biopesticides may be incorporated in tank-mix or rotational programs with conventional pesticides (to obtain additive effects or synergies, or as part of a resistance strategy), or used as standalone applications.

Biopesticides may be effective at moderate to severe disease pressures and can be as effective as conventional chemical pesticides.

Biopesticides can be cheaper than chemical pesticides when locally produced, and biopesticides are currently quicker and cheaper to register than conventional pesticides.

A key issue regarding the efficiency of biopesticides under field conditions is that they are affected by weather and pest pressure, and require considerable insight regarding mode of action, application timing and placement of the formulated product.

It is thus critical to first understand how a biopesticide product is designed to work in a given cropping program, and then to thoroughly evaluate the product based on the potential benefits.

Biopesticides can increase crop yields by working synergistically with chemical pesticides to optimize crop protection strategies, by allowing timely reuptake intervals and through their ability to be used close to harvest.

Since many biopesticides are slow to control the pest or pathogen, growers may feel that the product has insufficient effect and reapply unnecessarily.

Diligent reading of labels and understanding how the product works - their *mode of action* - is necessary for the optimal application of these products, allowing growers to reap the full benefits and attain consistent results.

Biopesticide demand is thus being driven by growers increased knowledge about the mechanisms of action (mode of action) of biopesticides and how their formulation affects the way we implement them.

I

PESTICIDE FORMULATION

1

INTRODUCTION TO PESTICIDE FORMULATION

Agrochemical formulation is the process of preparing a chemical product for use in agriculture. It involves the combination of active ingredients, such as pesticides, herbicides, and fertilizers, with various excipients and carriers to create a stable and effective product. While agrochemical formulation can be a highly effective way to protect crops and improve yields, it also presents a number of challenges that must be carefully considered and addressed.

One of the main challenges of agrochemical formulation is the stability of the active ingredients. Many agrochemicals are sensitive to temperature, humidity, and other environmental factors, which can cause them to break down or become less effective over time. Formulators must carefully select excipients and carriers that can help to protect the active ingredients from degradation and maintain their effectiveness.

Another challenge is ensuring that the formulated product is safe and effective for the intended use. This requires a thorough understanding of the properties and potential impacts of the active ingredients, as well as the requirements for application and handling. Formulators must also consider the potential for environmental impacts and the need for measures to mitigate any negative effects.

In addition to these technical challenges, agrochemical formulation also requires careful planning and coordination to ensure that the product can be produced efficiently and cost-effectively. This includes considerations such as sourcing raw materials, optimizing production processes, and ensuring compliance with regulatory requirements.

Overall, agrochemical formulation is a complex and multifaceted process that requires a deep understanding of the technical and regulatory challenges involved. By carefully addressing these challenges, formulators can create effective and safe products that help to support sustainable agriculture and meet the needs of farmers and other users.

2

BARRIERS TO ACTIVE INGREDIENT UPTAKE

One of the major challenges in agrochemical formulation is ensuring that the active ingredient is effectively absorbed by the target pest or disease.

This can be difficult due to various barriers that can prevent or reduce the uptake of the active ingredient. These barriers can be physical, chemical, or biological in nature, and can vary depending on the target organism and the environment.

Key barriers to active ingredient uptake include the cuticle or wax layer on the surface of the leaf or pest. In this section, we will discuss the various barriers to active ingredient uptake and strategies that can be used to overcome them in agrochemical formulation.

Bioavailability

Bioavailability refers to the extent an active ingredient becomes available to its intended biological target site.

Bioavailability can fall as much as two to three orders of magnitude before the active ingredient reaches the target site. The first order of magnitude loss (from 100% of the active ingredient in the bottle at purchase to 10% or less of the active ingredient reaching the target leaf or insect) is due to multiple (typically chemical, physical and environmental) factors.

These include precipitation and degradation in the bottle during storage and in the spray tank; drift," miss", and evaporation of the spray droplets between the spray nozzles and the target leaf.

Agrobiological factors largely determine bounce, run-off, volatility, wash-off, leaching, degradation and binding of the spray droplets and active ingredients to the leaf surface or soil.

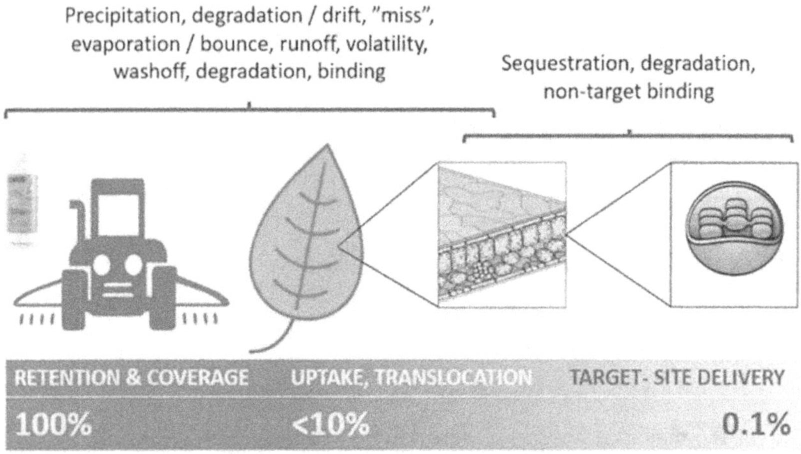

Figure 2.1: Bioavailability (the extent a substance or active ingredient becomes available to its intended biological target site) can fall as much as three orders of magnitude before reaching the target site.

An additional one to two orders of magnitude loss (from the 10% of the active ingredient reaching the target leaf to 1% or less reaching the target site) is due to several (typically agrobiological) factors, including sequestration of the active ingredient, metabolism and degradation, as well as general dilution and translocation within the plant.

From an agrobiological perspective, bioavailability is affected by the active ingredient concentration at the application site as well as barriers to active ingredient retention, coverage, uptake and translocation.

In turn, these barriers are determined by the anatomic and physiological characteristics of the absorptive site and the active ingredient's physical-chemical properties.

Barriers to active ingredient retention and coverage

For the agrobiologist, the first challenge to active ingredient retention and coverage is to hit the target crop or weed, and to ensure the spray droplets remain on the sprayed leaves.

Figure 2.2: Spray droplet loss by drift and evaporation, miss, shatter, bounce and run-off.

For this, an understanding of plant morphology (the physical form and external structure of plants), anatomy (the internal structure of plants) and physiology (plant function and behavior) is needed.

Flowering plants (crops and weeds) are divided into two main taxonomic groups:

- Monocotyledons (monocots) - grass and grass-like flowering plants the seeds of which typically contain only one embryonic leaf, or cotyledon. Much of the biomass in agriculture comes from monocotyledons, primarily grains (rice, wheat, maize).
- Dicotyledons, also known as dicots or broadleaves, with two embryonic leaves, or cotyledons

Of relevance for agrochemical formulation retention and coverage, several morphological differences are relevant:

- Leaf orientation (grasses tend to have an upright leaf orientation as opposed to the horizontal leaf orientation of broadleaves, presenting a smaller leaf area to hit and an increased risk of spray droplet bounce or run-off)
- Stomata (stomatal respiratory openings tend to be found on both ad- (upper) + abaxial (lower) leaf surfaces in grasses, as opposed to predominantly abaxial in broadleaves). Respiratory openings may be an entry route for pathogens as well as pesticides.
- Meristem (the single growing point or meristem in grasses tends to be protected at the base of the plant and thus challenging to target with herbicides, while broadleaves generally have multiple, exposed growing points)

- Roots (grasses tend to have fibrous roots with a large surface area near the soil surface, while broadleaves generally have tap roots that grow deep below the soil surface). These differences provide selectivity opportunities for pesticide application to root zones.
- Cuticle (grasses are generally more waxy and thus more challenging to wet than broadleaves).

Application parameters (droplet size, spray adjuvants, etc.) impact pesticide performance more on crops and weeds with difficult to wet leaves than on species with leaf surfaces that are easily wetted.

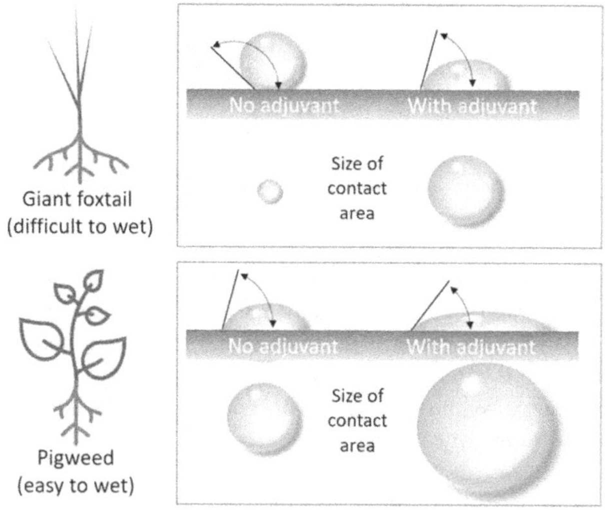

Figure 2.3: The crystalline wax platelets on foxtail leaves greatly reduce contact between the spray droplet and leaf, even with the addition of adjuvants. There is greater contact between the spray droplet and pigweed leaf surface than on foxtail.

Not all grasses are difficult to wet – examples of easy-to-wet grasses include members of the Bromus family and volunteer cereals. In contrast, some broadleaves, such as members of the Brassica family, Cleavers, Goosefoot etc., can be difficult to wet.

Finally, spray droplets may "pearl" and remain suspended on leaf hairs, called trichomes (e.g., soya) without contacting the leaf; in the absence of trichomes (e.g., cotton) the spray droplets spread and cover the surface of the leaf.

In addition to differences in leaf surface characteristics among species, the amount and kinds of epicuticular cuticular wax depend on environmental conditions. Under dry conditions, plants produce more cuticular wax to limit water loss through evapotranspiration. The cuticle may swell or shrink depending on humidity and the moisture status of the plant, increasing or decreasing the permeability of active ingredient molecules.

Foliar barriers to active ingredient uptake

The cuticle consists of an upper epicuticular wax layer covering a cutin and wax matrix. Wax fatty acids are responsible for the negative charge of the leaf. Cationic active ingredients may bind to the negatively charged leave surfaces, making them unavailable for uptake by the plant. During foliar uptake, active ingredients must move from lipophilic to hydrophilic environments.

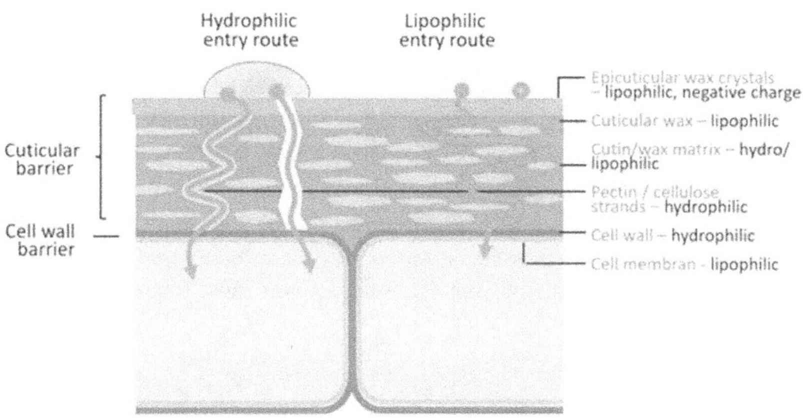

Figure 2.4: Active ingredient uptake - foliar

Lipophilic pesticide molecules easily diffuse and penetrate into the lipophilic cuticular wax. Uptake of lipophilic compounds across the cuticular barrier (lipophilic or non-polar entry route) is facilitated through molecular-scale spaces that form temporarily by the movement of amorphous cutin and wax fractions.

While made up predominantly of lipophilic waxes, the cuticular barrier also contains hydrophilic pathways comprising cellulose, pectin and cutin, which serve as polar pathways for active ingredient uptake into the leaf (hydrophilic or polar entry route). In addition, molecular scale cracks form across the cuticular barrier, allowing the passage of water and water-soluble compounds.

The cell wall, a mixture of cellulose, hemicellulose and other hydrophilic substances, and the lipophilic cell membrane is the final barrier to active ingredient uptake.

To facilitate the retention and uptake of hydrophilic and lipophilic active ingredients across the cuticular barrier, formulation scientists add specific adjuvants (such as stickers, spreaders or penetrators) to the formulation.

Root barriers to active ingredient uptake

For soil-applied pesticides, active ingredient uptake primarily takes place via the roots. Root epidermal cells have root hairs which increase the surface area of the root for water absorption.

As roots are not covered by a lipophilic cuticle, formulations for soil-applied pesticides generally do not require the inclusion of adjuvants such as spreaders and penetrators.

The uptake of active ingredients into roots can take place via symplastic (intracellular) and apoplastic (extracellular) pathways,

In the symplastic uptake pathway, water containing the active ingredient moves through plasmodesmata, openings between cells which permit the flow of small, solubilized molecules.

The apoplastic pathway allows active ingredient uptake through the cell walls by capillarity until it reaches the endodermis, where the lipophilic Casparian strip blocks their passage, and the active ingredient must transition into the symplast.

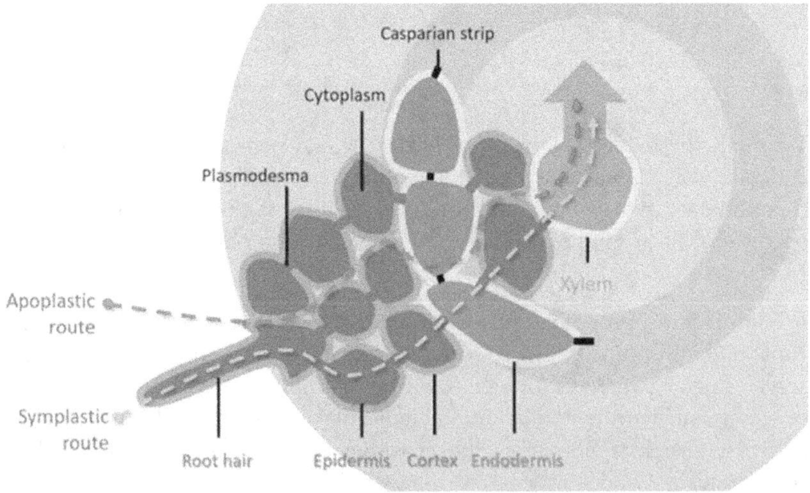

Figure 2.5: Active ingredient uptake & transport in roots.

Through these pathways, soil-applied active ingredients are moved into the xylem for translocation into the plant.

Active ingredient translocation

Active ingredient polarity determines active ingredient uptake and translocation via apoplastic (extracellular), symplastic (intracellular) pathways,

Many pesticide active ingredients are weak acids, and can accept or donate hydrogen ions depending upon the pH of their environment. The ability of weak acids to change their polarity depending upon pH is a critical factor in the uptake and translocation process.

pH gradients facilitate active ingredient translocation within the plant. Plants have specialized carrier proteins that pump hydrogen ions from the cytoplasm across the cell membrane into the intercellular spaces. This raises the cytoplasm's pH (more alkaline), whereas it lowers the pH (more acid) outside the cell.

Weak acids become protonated in the acidic environment outside the cell (gain hydrogen ions from the solution). They become more lipophilic, facilitating uptake through the epicuticular wax or the cell membrane, apoplastic movement between cells and translocation within the plant via the moderately acidic xylem vascular system.

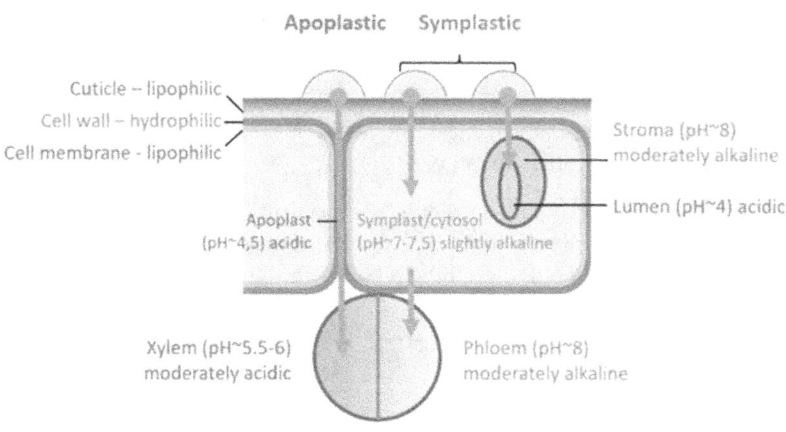

Figure 2.6. Apoplastic (extracellular) and symplastic (intracellular) active ingredient uptake and transport pathways

In the alkaline and aqueous cytoplasm, the active ingredient becomes deprotonated (donates hydrogen ions to the solution and acquires a negative charge). It becomes more water soluble, facilitating intracellular symplastic movement and translocation within the plant via the moderately alkaline phloem vascular system.

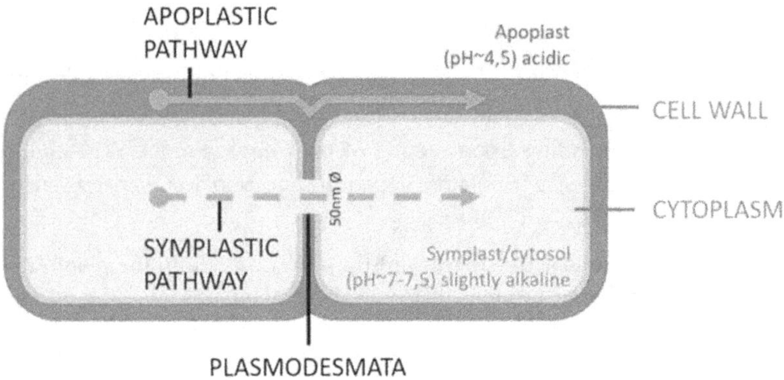

Figure 2.7. Apoplastic (extracellular) and symplastic (intracellular) active ingredient transport between cells

Plasmodesmata are approximately 50 nm in diameter, although larger molecules can also be transported. For transport between cells, plasmodesmata allow the direct flow of small molecules such as sugars, amino acids, and ions between cells.

This opens the possibility of developing stimuli-responsive and targeted delivery nanocarriers that can mask the active ingredients' physical-chemical properties and deliver them to target sites.

To predict pesticide mobility in plants, we need to consider the dissociation constant of the molecule (pKa), i.e., the charge of the molecules at physiological pH and LogP (lipophilicity) of neutral (uncharged) molecules.

3

ACTIVE INGREDIENT UPTAKE AND MOBILITY

The physical and chemical properties of the active ingredient play a crucial role in the performance and effectiveness of an agrochemical formulation. These properties can affect the solubility, stability, and bioavailability of the active ingredient, as well as the physical and chemical properties of the formulation itself.

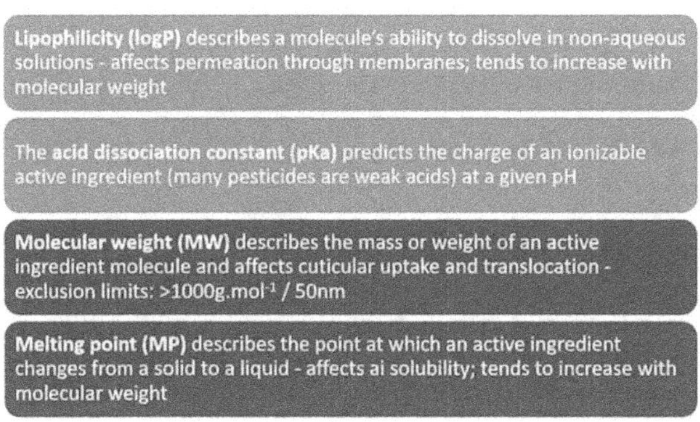

Figure 3.1 Key physical-chemical characteristics for the penetration of active ingredients into plant leaves.

Some important physical-chemical properties of active ingredients that are relevant to agrochemical formulation include molecular weight, solubility, melting point, vapor pressure, and pKa.

In this section, we will discuss the physical-chemical properties of active ingredients and their relevance to agrochemical formulation. We will also explore the various factors that can influence these properties and the strategies that can be used to optimize them for specific applications.

Depending on its intended use, a pesticide active ingredient (a.i.) may require an element of water solubility (hydrophilicity) to allow it to be transported and distributed within the plant, while having some non-aqueous solubility (lipophilicity), to allow it to permeate across biological membranes.

The penetration of active ingredients into plant leaves is thus related to the physicochemical properties of the active ingredients, primarily lipophilicity (logP), dissociation (pKa), and secondarily molecular weight (MW) and melting point (MP).

Molecular Weight (MW)

The diffusion rate of lipophilic and hydrophilic compounds across leaf cuticles negatively correlates with their molecular weight (MW). The plasmodesmata of plants can only allow free circulation of molecules smaller than a molecular weight of 1000 (g/mol).

Within the 100–500 molecular weight range typical of pesticide active ingredients, the effect of molecular size on pesticide uptake into plant leaves may not be of great importance, as molecular weight is not the sole factor affecting chemical uptake.

Melting point (MP)

The melting point (the temperature at which crystals are in equilibrium with the liquid phase at atmospheric pressure) affects the solubility of the active ingredient and thus the bioavailability of the substance i.e. the ability to be transported to the active site. The melting point increases with molecular size because the molecular surface area available for contact with other molecules increases.

Pesticide-likeness rules of thumb

Within (pharmaceutical drug discovery, Lipinski's rule of five is a rule of thumb to evaluate if a molecule's physical-chemical properties make it likely to be an active drug.

The rule was formulated on the observation that most orally administered drugs are relatively small and moderately lipophilic molecules, and suggests that an orally active drug has:

- Lipophilicity (logP) not greater than 5 (not too lipophilic)
- No more than 5 hydrogen bond donors (not too polar). Hydrogen-bond donors and acceptors are involved in weak non-chemical bond interactions, and their number and ratios are important in binding, influencing the pKa of compounds, interacting with active sites, etc.
- A molecular mass of less than 500 g/mol (not too big)

Lipinski's rule are, however, not a suitable screening guide for pesticide leads, and alternative rules for leads of fungicides, insecticides, and herbicides have been proposed. The key rules are outlined here:

	Lipinski Rule of 5	Brigg's Rule of 3	Tice's Rules	
	Drugs	Fungicides	Herbicides	Insecticides
logP	≤ 5	≤ 3	≤ 3.5	0-5
H-bond donors	≤ 5	≤ 3	≤ 3	≤ 2
H-bond acceptors	≤ 10	-	2-12	1-8
Melting point	-	≤ 300	-	-
Molecular weight*	≤ 500	~300	150-500	

*within the 100–500 molecular weight range, the effect of molecular size on ai uptake and translocation is of less importance, than logP and pKa

C.A. Lipinski, F. Lombardo, B.W. Dominy, P.J. Feeney, Advanced Drug Delivery Reviews 1997, 23, 3..
G.G. Briggs, 'Predicting uptake and movement of agrochemicals from physical properties' talk given at the SCI Meeting on uptake of agrochemicals and pharmaceuticals, London, UK, December,1997
C.M. Tice, Pest. Manag. Sci. 2001, 57,

Figure 3.2: Lipinski's rule for drug-likeness and Brigg's and Tice's rules for leads of fungicides, insecticides, and herbicides.

Lipophilicity (logP)

Lipophilicity describes a molecule's ability to dissolve in lipophilic (non-aqueous) solutions, allowing permeation through biological membranes.

Lipophilicity is measured as the partition coefficient (Kow) of a given molecule between non-aqueous (octanol, o) and aqueous (water, w) phases, and is expressed as the logarithm of the ratios between these phases (logKow or logP).

A logP value of 3 indicates a distribution of one molecule in the aqueous phase for every 1000 molecules in the lipophilic (octanol) phase – the molecule may be considered to be lipophilic.

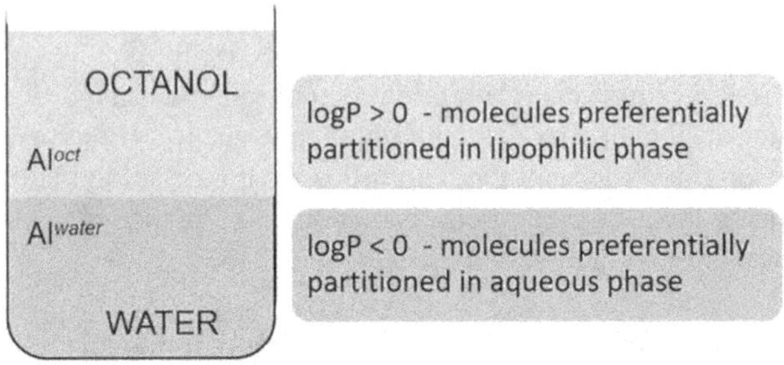

Figure 3.3: Partitioning of an active ingredient (AI) molecule between non-aqueous (octanol, o) and aqueous (water, w) phases.

For logP=0, the molecules will be equally partitioned between the two phases, while for logP values < 0 the molecules will be preferentially partitioned in the aqueous phase. LogP only describes the partition coefficient of neutral (uncharged) molecules.

To predict pesticide mobility in plants, it is necessary to consider the dissociation constant of the molecule, i.e. the charge of the molecules at physiological pH.

In contrast, the distribution coefficient (logD) is measured at different pH levels, and thus considers the charge of the active ingredients molecules.

pKa

The acid dissociation constant (Ka) describes the strength of an acid and is typically expressed as pKa (-log10(Ka)). The pKa value of a given molecule defines the pH at which it is neutral.

At greater pH values, acid groups will be charged, while at lower pH values basic groups will be charged. The number and distribution of charges on a molecule affect its aqueous solubility.

Fluids in plants generally have narrow pH ranges. Phloem sap is alkaline with a pH of 8-8.5 while xylem and apoplastic (extracellular) solutions are acidic with a pH of 5-6. Within plant cells the cytoplasm is slightly alkaline (pH 7.4-7.5) while the vacuole is acidic with a pH of 4.5-6.

In addition to the pH- and pKa-dependent effect on the overall charge of the active ingredient, trans-membrane pH gradients are able to drive the transport of some molecules across biological membranes through membrane pores.

• • •

Pesticide Solubility and Uptake

Solubility of the active ingredient is determined by its lipophilicity and dissociation constant. Lipophilic active ingredient molecules may resist solubility in water – the most common medium for agricultural sprays – and typically require formulation with organic solvents or oils.

Once on the insect or leaf surface, the lipophilicity of these molecules allows the rapid uptake of the active ingredient across lipophilic waxy, cuticular and chitin surface barriers, as well as facilitating extracellular movement within biological cell membranes and cell walls.

In contrast, hydrophilic active ingredient molecules are easily solubilized in aqueous spray solutions, while the uptake of these molecules across lipophilic plant leaf barriers often requires the inclusion of surfactants in the final product.

Delayed or insufficient uptake of the active ingredient may lead to reduced efficacy, wash-off of the active ingredient by rain, or UV-degradation of the unprotected molecules.

Translocation

Within the plant, active ingredient molecules may be translocated throughout the plant according to their lipophilicity and dissociation constants.

If the active ingredient remains on the plant surface, it is described as having contact activity, whereas if the active ingredient is translocated throughout the plant, it is described as having systemic activity.

Contact pesticides tend to be lipophilic, and are absorbed into their waxy cuticle of plants, where they resist wash-off by rain. Contact insecticides are particularly effective against so-called rasping insects, such as mites, which feed on surface cells.

Alternatively, lipophilic insecticides may be applied to the soil, where their lipophilicity contributes to reduced leaching of the active ingredients.

Systemic pesticides tend to be hydrophilic, and – depending on the logP and pKa of the active ingredient – translocation of systemic pesticides may take place via the phloem or xylem transport tissues in vascular plants.

The phloem transports sap bidirectionally, from existing leaves upward in the plant to new developing tissue, as well as downwards to the roots, to which glucose formed in leaves may be translocated and stored in the form of sucrose.

In contrast, the function of the xylem is to transport water unidirectionally from roots to shoots.

Thus, systemic active ingredients transported via the phloem may be translocated throughout the plant, including to leaves not otherwise reached by these spray solution, as well as to the roots (of relevance for insecticidal active ingredients).

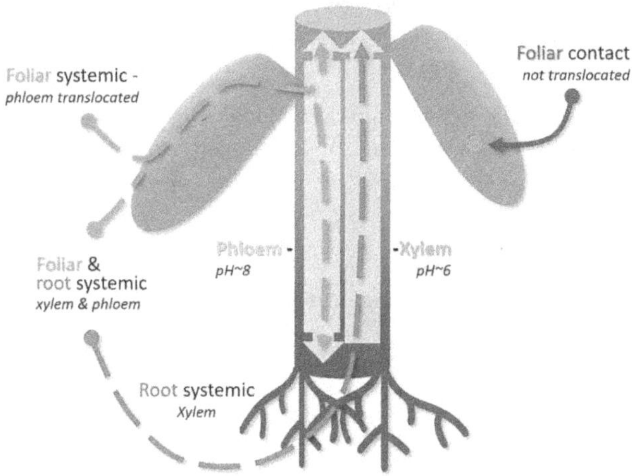

Figure 3.4 Phloem and xylem translocation in vascular plants.

Active ingredients transported by the phloem are ingested by sucking insects such as aphids, which preferentially feed on the nutrient-rich phloem.

Systemic active ingredients transported via the xylem and are translocated "upwards" in the plant, from lower leaves to shoots, but not to the roots.

Interestingly, while systemic activity is generally correlated with increased water solubility (< logP 3.2), translaminar activity is reported for both water-soluble (e.g., the systemic insecticide Pymetrozine; logP -018) as well as lipophilic ai's (e.g., Abamectin; logP 4.4).

This may be dependent on whether the translaminar pathway is extracellular (lipophilic environment or intracellular (hydrophilic environment).

Some insecticidal active ingredients show both systemic as well as translaminar activity, predominantly hydrophilic molecules such as Acephate (logP -0,89), Pymetrozine (logP -0.18) and Thiamethoxam (logP -0.13, as well as Imidacloprid (logP o.57) and Cyantraniliprole (logP 2.02).

For insecticide efficacy testing, the degree (or lack) of systemic and translaminar activity will determine the active ingredient uptake pathway, and thus the pest panel needed to cover contact, systemicity, translaminar ai pathways.

Bromilow Model - Solubility and Pesticide Mobility

The Bromilow model was developed and presented by RH Bromilow et al. in the early 1990s, and considers the partition coefficient (logP) while taking the charge of the active ingredient molecules into account.

LogP and pKa are plotted against each other, and the resulting diagram is a useful model for the prediction of pesticide mobility in plants at physiological pH.

In the model, pesticide mobility is classified as phloem- & xylem-mobile, xylem-mobile or non-mobile. Depending on the logP and pKa of the pesticide molecule, a pesticide may be classified as non-mobile (contact) or mobile (systemic).

Pesticide active ingredients with a logP value of less than 3.2 are considered systemic, with hydrophilicity and systemicity increasing with decreasing logP values.

Decreasing pKa is associated with increased dissociation of the pesticide active ingredient, with a concomitant increase in plant systemicity.

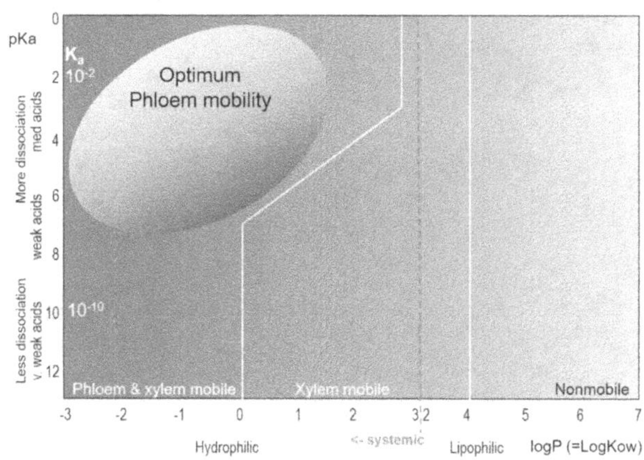

Figure 3.5: Bromilow model for the prediction of pesticide mobility in plants (degrees of mobility as a function of logP and pKa), adapted from Bromilow et al. (1991).

Where no pKa value is given for an active ingredient, it may indicate that no significant dissociation occurs in the environmental range; values of pKa <4 for bases and pKa >8 for acids are not relevant in an environmental context.

• • •

Bromilow Model – Hydrophilic vs. Lipophilic

Glyphosate (GLY) is a non-selective, water-soluble herbicide with a logP of -2.8 and a dissociation constant of about pKa≈2.34 (glyphosate exists as a series of zwitterions, and therefore exists in different ionic states depending on pH), placing it into the optimum phloem mobility category.

Glyphosate is readily translocated throughout the plant via the phloem and xylem, moving from sprayed leaves to new shoots, where it inhibits the catalytic function of the EPSP (5-endopyruvylshikimate-3 phosphate synthase) enzyme, thereby inhibiting the formation of aromatic amino acids necessary for plant growth.

In contrast, Diflufenican (DFF) is a lipophilic herbicide (logP=4.2) used to control grasses and broadleaf weeds, with a non-systemic mode of action. DFF is non-mobile within plants, and its herbicidal activity is restricted to the immediate vicinity of the application droplets. DFF is used as a pre-emergent herbicide, and is typically sprayed directly to the soil surface, where it forms a stable, lipophilic herbicide barrier.

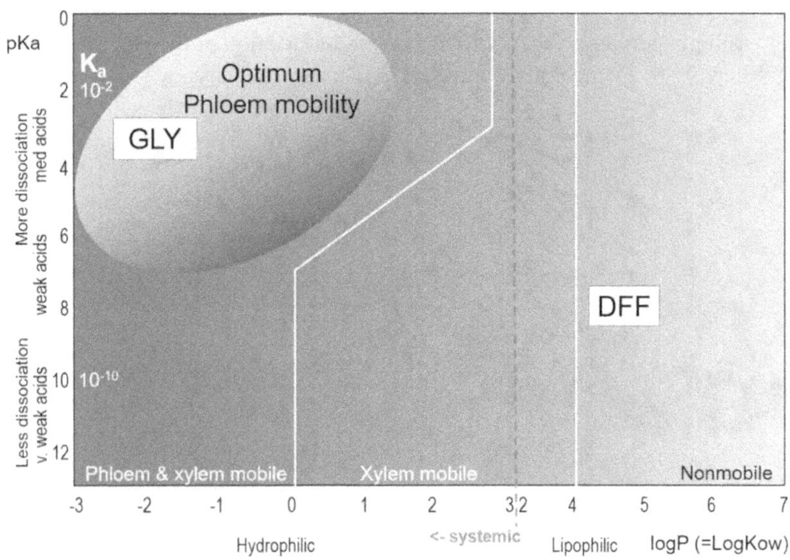

Figure 3.6: Bromilow model for the prediction of herbicide mobility in plants of Glyphosate (GLY) and Diflufenican (DFF).

Germinating weed shoots encounter the active ingredient as they break through the soil surface, which causes bleaching by blocking the production of protective plant antioxidants through its inhibitory effect on the enzyme PDS (phytoene desaturase), ultimately leading to death of the seedling through oxidative damage.

Bromilow Model – Pesticide mobility

By including herbicide, fungicide and insecticide active ingredients in the Bromilow model, we can determine and visualize pesticide mobility for a broad range of active ingredients.

In Figure 3.7, we see that many common herbicides tend towards xylem and phloem mobility; they are systemic and are therefore typically applied as post-emergent (Post em.) foliar treatments.

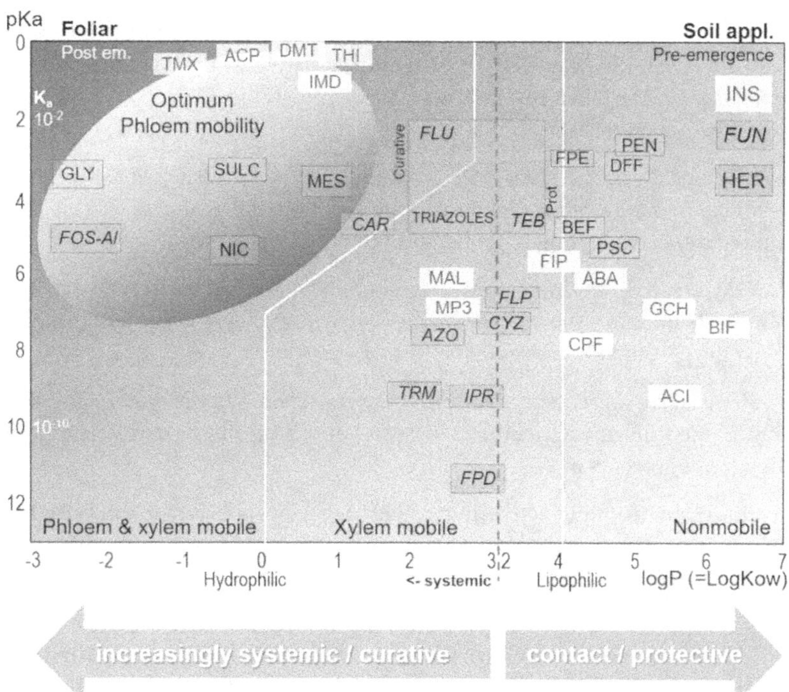

Figure 3.7: Bromilow model for the prediction of pesticide mobility in plants for insecticides (INS), fungicides (FUN) & herbicides (HER). **ABBREVIATIONS - Insecticides**: TMX thiamethoxam ACP acetamiprid DMT dimethoate IMD imidacloprid THI thiacloprid MAL malathion MP3 methyl parathion FIP fipronil ABA abamectin CPF chlorpyrifos GCH gamma-cyhalothrin ACI acrinathrin BIF bifenthrin. **Fungicides**: FOS-Al fosetyl-aluminum CAR carfentrazone FLU flutriafol TEB tebuconazole FLP fluazifop CYZ cyazofamid AZO azoxystrobin TRM thiram IPR iprodione FPD fenpropidin. **Herbicides**: GLY glyphosate SULC sulcotrione NIC nicosulfuron MES Mesotrione FPE fenoxaprop-p-ethyl BEF beflubutamid DFF Diflufenican PEN pendimethalin PSC prosulfocarb.

Other herbicides such as DFF and Pendimethalin (PEN; logP=5.4, pKa=2.8) – an inhibitor of cell division and thus root and shoot growth – are non-systemic, and are applied as a pre-emergence (or early post-emergence) soil applications.

Insecticides may show great systemicity within the plant e.g. Imidacloprid (IMD; logP=0.57) making them ideal to combat sucking pests such as aphids, or non-mobility, e.g. Bifenthrin (BIF; logP=6.6) making them suitable for the control of rasping, surface-feeding pests such as mites and thrips.

Many common fungicides can be seen to be clustered around the "low" end of the systemicity scale, and are primarily xylem mobile. For triazoles, systemicity may be relatively low e.g. Tebuconazole, TEB; logP=3.7, pKa=5.0) while Flutriafol (FLU; logP=2.3, pKa=2.3) is the most systemic triazole.

Here, the Bromilow model demonstrates its value for plant physiologists: although the Triazoles lie within a relatively narrow interval in the Bromilow model, their mobility ranges from systemic to relatively non-systemic, and care should be taken before generally recommending Triazole fungicides as curative treatments.

For a fungicide to be considered curative, it is necessary for the active ingredient to be systemically translocated within the plant, to enable it to inhibit (cure) existing fungal hyphal growth within the leaf.

In contrast, protective fungicides are generally described as having a contact action, remaining on the surface of the leaf where they can protect the plant against the germination of newly landed fungal spores.

Lipophilic active ingredient molecules may resist solubility in water – the most common medium for agricultural sprays – and typically require formulation with organic solvents or oils.

In contrast, hydrophilic active ingredient molecules are easily solubilized in aqueous spray solutions, while the uptake of these molecules across lipophilic plant leaf barriers often requires the inclusion of surfactants in the final product.

Delay or insufficient uptake of the active ingredient may lead to reduced efficacy, wash-off of the active ingredient by rain or UV-degradation of the unprotected molecules.

Propesticides

A propesticide is a pesticidally inactive molecule requiring transformation after its application to become pesticidally active. The modification of propesticides influences the biological activity of the parent molecule, including:

- Increased solubility or lipophilicity, influencing distribution(translocation) within the target organism
- Increased bioavailability and target selectivity
- Reduced toxicity towards non–target species through differential metabolism
- Controlled biokinetics, through delayed release of the active metabolite from its parent propesticide structure

Propesticide metabolism is species-dependent, due to differences in enzyme activity required for their activation. Propesticides include:

- *Proinsecticides*: neural inhibitors Chlorpyrifos, Thiamethoxam, and Indoxacarb; ACCase inhibitor Spirotetramate, mitochondrial inhibitor Chlorpenafyr and bacterial protein delta-endotoxin
- *Profungicides*: Prothioconazole, Mancozeb, Thiophanate-Methyl, Dazomet, and Fosetyl-Aluminum
- *Proherbicides*: the ACCase inhibitors Pinoxaden, Clethodim and Fenoxaprop; protox inhibitors Saflufenacil, Flumioxazin, and Carfentrazone-ethyl), as well as Fluroxypyr, Isoxaflutole, and Clomazone

The proinsecticide Chlorfenapyr is metabolized to its active form within the insect. Although the metabolite is toxic to mammals, these lack the metabolic enzymes to make the conversion. This differential metabolism provides reduced toxicity towards non–target species.

The profungicide Prothioconazole (logP=4.05; pKa=6.9) is metabolized to the fungicidally-active prothioconazole-desthio, (logP=2.42).

Thus, while Prothioconazole is considered to be systemic, the systemicity is actually due to its conversion to prothioconazole-desthio:

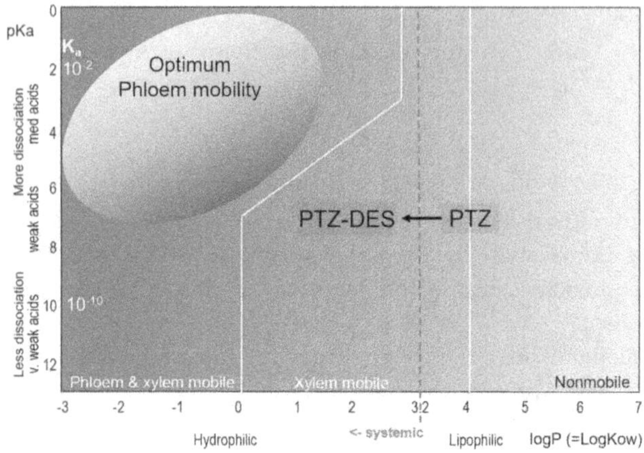

Figure 3.8. Bromilow model for the prediction of profungicide mobility in plants of Prothioconazole (PTZ) and Prothioconazole-desthio (PTZ-DES).

When evaluating the translocation of a molecule within a plant, it may be necessary to determine whether the active ingredient is metabolized immediately after uptake.

For example, the grass-weed herbicide Fenoxaprop-p-ethyl (FPE; logP=4.58; pKa=0.18) is formulated as an ester, which facilitates its uptake across lipophilic plant barriers.

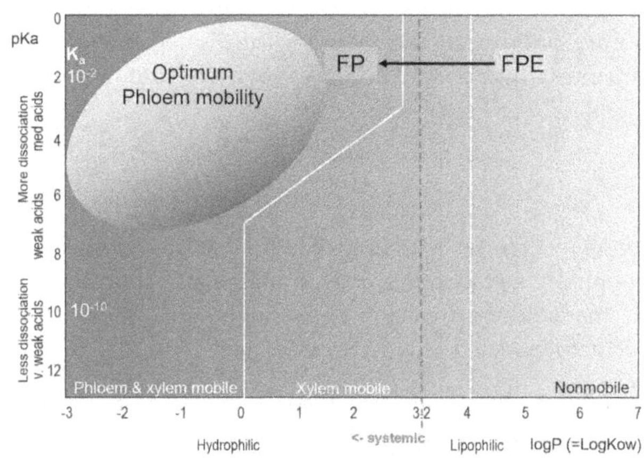

Figure 3.9. Bromilow model for the prediction of proherbicide mobility in plants of Fenoxaprop-p-ethyl (FPE) and Fenoxaprop-p (FP).

Within the plant, however, the molecule is de-esterified to its toxic form, and the resulting Fenoxaprop-p molecule (FP; logP=1.5) exhibits herbicidal activity as well as increased hydrophilicity, facilitating translocation through the plant via both the xylem and phloem.

4

FORMULATION ADJUVANTS

Adjuvants are additives that are added to agrochemical formulations to improve their performance or to modify their properties.

They can be used to enhance the efficacy of the active ingredient, to improve the physical properties of the formulation, or to modify the way in which the active ingredient is released or applied.

Adjuvants can be classified into several categories, including surfactants, emulsifiers, spreading agents, wetting agents, and sticker agents.

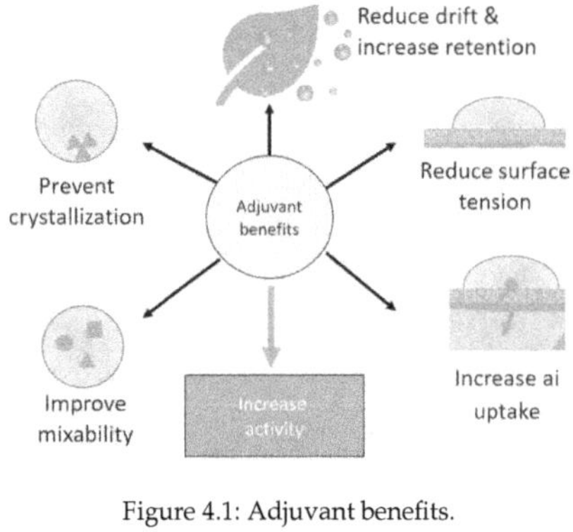

Figure 4.1: Adjuvant benefits.

Each type of adjuvant has specific properties and functions that are designed to improve the performance of the agrochemical formulation. In this section, we will discuss the various types of adjuvants, their properties, and their uses in agrochemical formulation.

Adjuvants fall into two groups: Activator adjuvants (improving the activity of the pesticide) and Modifier or Utility adjuvants (those improving the handling and application of the pesticide).

Together, adjuvants modify the spray solution to improve handling, application and efficacy in the case of conventional pesticides.

Activator Adjuvants

Activator adjuvants command the greatest market share, and their function is to modify spray mixture properties to optimize pesticide and biopesticide retention and uptake by improving their spreading and penetrating properties, thereby improving the efficacy of the product.

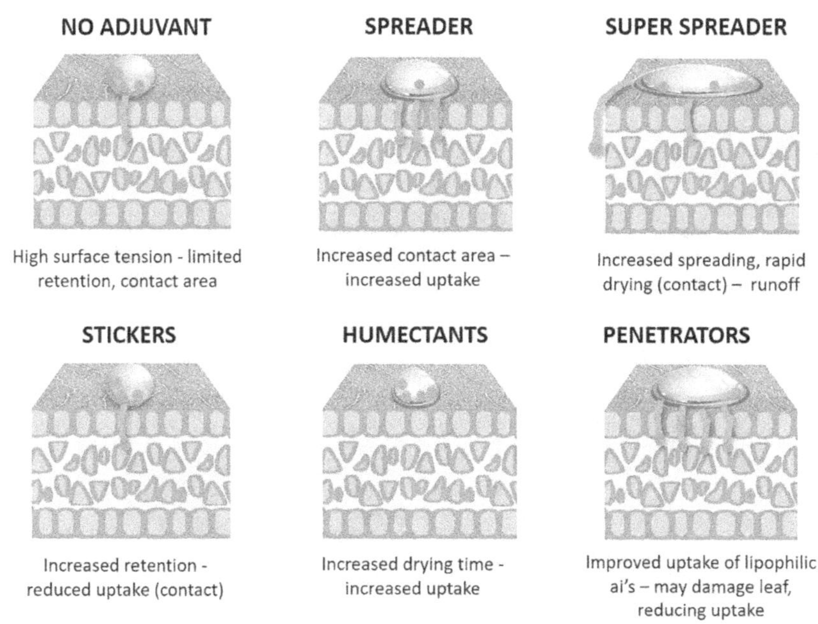

Figure 4.2: Overview of activator adjuvants.

Wetters / spreaders

Surfactants – or surface acting agents – are most commonly used to increase spray coverage on leaf surfaces, by reducing the surface tension of spray droplets and allowing them to spread beyond their initial contact area.

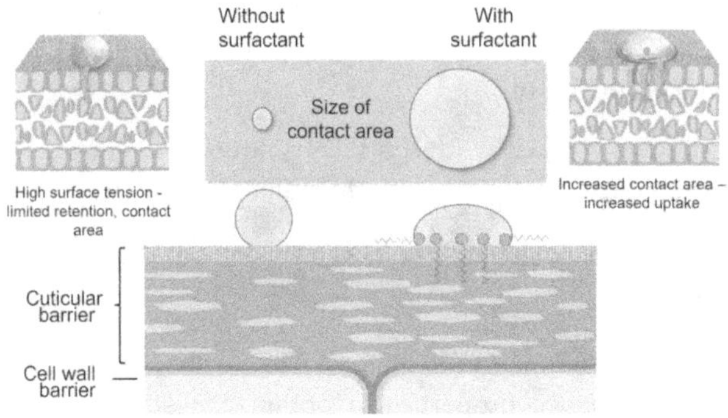

Figure 4.3: Surfactants increase spray overage on leaves.

Surfactants adsorb at the interface between the cuticle and the spray droplets, where the lipophilic tails of the surfactant molecules favour the lipophilic cuticle, while the polar hydrophobic "heads" of the surfactant molecules favour the water phase.

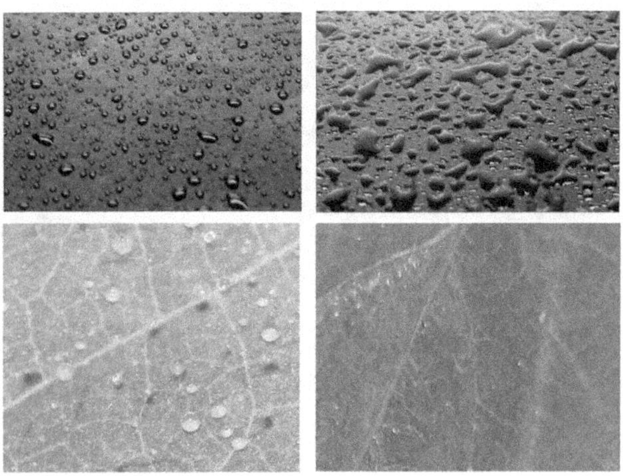

Figure 4.4: Spray droplets without adjuvants (top left) and with added wetter adjuvant (top right). In the absence of adjuvants, spray droplets "pearl" and can be seen to sit on leaf hairs (bottom left) without contacting the leaf; with the addition of a super spreader, the spray drops spread and cover the entire surface of the leaf.

The *hydrophilic-lipophilic balance* (HLB) of surfactants is a measure of their hydrophilicity or lipophilicity (ratio of weight percentages of hydrophilic to hydrophobic groups in the structure). The HLB system allows us to predict how surfactants will behave in pesticide formulations.

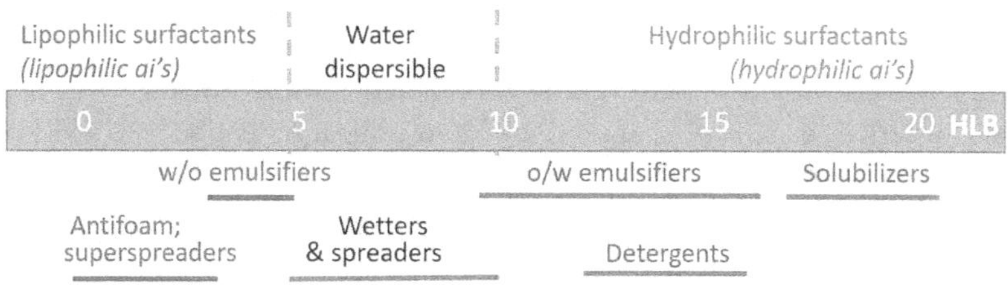

Figure 4.5: HLB and lipophilic, water-dispersible and hydrophilic surfactants.

Lipophilic surfactants in the HLB range of 1 to 10 generally work best with oil-soluble active ingredients, and hydrophilic surfactants in the HLB range of 10 to 20 generally work best with water-soluble active ingredients. Lipophilic surfactants include antifoaming agents and water-in-oil emulsifiers. Surfactants with HLB values of less than 2 are classified as spreading agents. Spreading agents are insoluble in water and do not emulsify or solubilize lipids or membranes.

Hydrophilic surfactants (HLB>10) include oil-in-water emulsifying agents, detergents and solubilizing agents. Common detergents have HLB values greater than ten. They adhere to lipids and influence the interaction of soluble macromolecules

Water-dispersible surfactants (HLB 6-10) include oil-in-water emulsifiers and wetting and spreading agents.

Surfactants may be charged (non-ionic) or charged (anionic or cationic). Some of the most common non-ionic surfactants are EO/PO copolymers in which the hydrophilic part is EO (ethylene oxide) and the hydrophobic part is PO (propylene oxide). Shorter EO chains are generally used for lipophilic active ingredients, while longer EO chains are used for hydrophilic active ingredients.

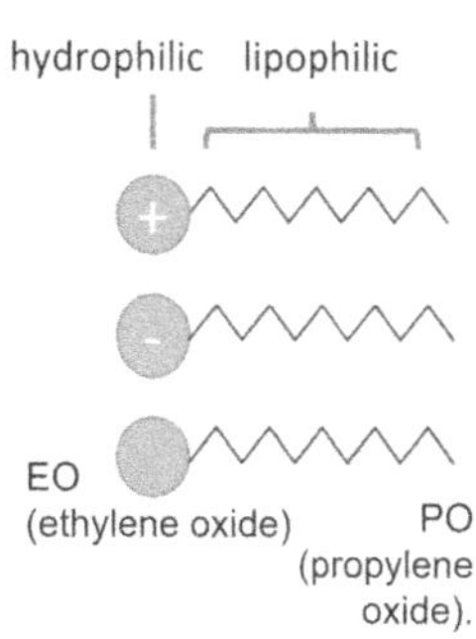

Figure 4.6: Charged and uncharged surfactants.

Anionic surfactants are typically used to improve the efficiency of contact insecticides to control small, relatively immobile pests such as mites, thrips and aphids. A few dry (solid) formulations utilize anionic surfactants (net negative electrical charge) built into the formulation to assist in dispersibility when added to the spray tank.

Cationic surfactants (with a net positive electrical charge) are occasionally applied as agricultural surfactants but are rarely used as they tend to be phytotoxic.

Non-ionic (having no net electrical charge) surfactants (**NIS**) are widely used in agricultural applications, either added to the pesticide formulation, supplied as a stand-alone adjuvant or included in an adjuvant blend.

Non-ionic surfactants have the advantage to ionic surfactants in that one can obtain surfactants with wide range of hydrophile-lipophile balance (HLB) by changing molecular structures, especially hydrophilic moieties (e.g. ethylene oxide, EO), with shorter EO chains for lipophilic active ingredients and longer EO chains for hydrophilic active ingredients. NIS are compatible with most pesticides and are typically used with *systemic* products.

Recently, surfactants based on a trisiloxane-backbone (**organosilicones**) have been developed for agricultural applications. Organosilicons are silicon-based surfactants with extreme spreading ability ("super-spreaders"). Super spreaders can facilitate the movement of pesticides deep into the crop canopy, or into the other difficult-to-reach areas such as leaf whorls.

Examples of common surfactants by type

Anionic surfactants	Cationic Surfactants	Nonionic surfactants (NIS)				Organosilicones
Contact insecticides; dry formulations (dispersibility)	Rarely used (phytotoxicity)	Spreaders, wetter & stickers – compatible with most pesticides (all-purpose) and are typically used with *systemic* products				Superspreaders, penetrants (stoma), rainfastness (but increased evaporation and run-off)
Alkyl sulfates: sodium lauryl sulfate (sodium dodecyl sulfate, SLS, or SDS)	**Quaternary ammonium salts**	**Ethoxylates:** Triton X-100, Polyethoxylated tallow amine	**Fatty acid esters:** Tween 20, Tween 80	**Alcohol ethoxylate** "wetters"	**Alkylaryl ethoxylate** (nonyl & octylphenol) "wetters"	**Organosilicones:** straight or plus NIS; plus latex

Nitrogen-surfactant blends decrease surface tension, increase spreading of herbicides on the leaf surface, neutralize ionic charges, and increase herbicide penetration into the leaf

Figure 4.7: Overview of common surfactants by type.

While super spreaders can offer enhanced coverage and increased penetration through stomata, their use may lead to increased rates of spray droplet evaporation (with potentially lower pesticide uptake rates), or coalescence and lead to run-off from the leaf surface.

Penetrants / translocators are generally used to optimize active ingredient uptake in the grasses as well as broadleaf plants with a thick cuticle.

Plants growing under hot, dry conditions may develop a thicker cuticle to reduce desiccation, and the inclusion of penetrants in spray solutions may be particularly beneficial under these conditions.

By affecting uptake across the cuticle as well as intracellular or extracellular transport within the plant, adjuvants are an additional factor deciding whether chemical and biochemical active ingredients have apoplastic (extracellular), symplastic (intracellular) or translaminar (from adaxial to abaxial leaf surface) activity.

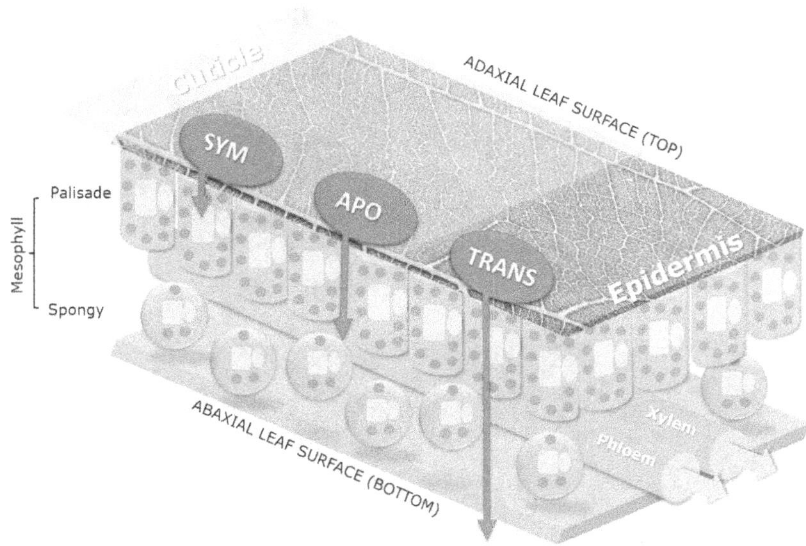

Figure 4.8: apoplastic (extracellular), symplastic (intracellular) or translaminar (from adaxial to abaxial leaf surface) active ingredient pathways.

Crop oil concentrates (COC; petroleum or vegetable-based oil with surfactants to emulsify the oil) and methylated seed oils (MSO; modified vegetable-based oil) function as penetrants (facilitating active ingredient uptake by softening or dissolving leaf cuticular waxes) and translocators (facilitating movement within cell walls and membranes), and may act as humectants.

Nitrogen-surfactant blends consist of premix combinations of various forms of nitrogen (e.g. Ammonium fertilizers) and surfactants, and are generally used with herbicides.

The surfactants reduce surface tension and improve leaf surface spreading, while the nitrogen compounds neutralize ionic charges, prevent active ingredient precipitation in the tank mix or on the leaf surface, and promote herbicide uptake primarily into broadleaf weeds.

Modifier (Utility) Adjuvants

Modifier adjuvants affect the physical properties of the formulated product in the bottle or in dilution in the spray tank.

Modifier adjuvants may be grouped into those affecting the physical properties of the spray solution (spray modifier adjuvants) and those affecting their handling and application (utility modifier adjuvants).

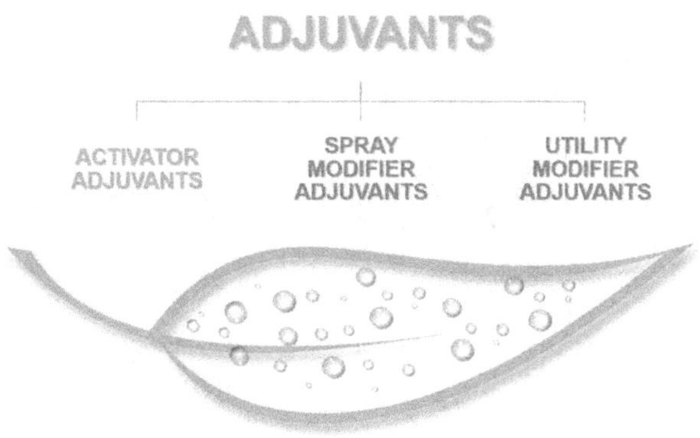

Figure 4.9: Adjuvant groups.

Modifier adjuvants address specific conditions that may affect the spray solution or its application, typically by modifying the physical characteristics of the spray solution, facilitating handling, safety and application.

In-can adjuvant refers to primarily modifier adjuvants incorporated into pesticide formulations. When these are added separately when the spray solution is prepares, the adjuvant is called a tank mix adjuvant.

• • •

Spray Modifiers

Spray modifiers affect the physical properties of the spray solution and include stickers, drift retardants (thickeners, typically polymers which minimize small droplet formation) and humectants.

Stickers help spray droplets adhere to the target, and subsequently reduce run-off and wash-off by rain, by increasing the adhesion or "stickiness" of active ingredients, mainly when applied to difficult-to-wet species (brassica; cereal crops and grass weeds)

Many stickers contain surfactants and are termed **spreader-stickers**, which give both a sticker action and a wetter-spreader action. Surfactants reduce droplet surface tension while the fatty acid component of the sticker increases the adhesion of spray drops to the leaf and reduces run-off / wash-off.

The drying time of spray droplets determines the duration of optimal active ingredient uptake, through extended suspension or solubility of the active ingredient.

Humectants are hygroscopic formulation additives that prevent premature spray droplet desiccation, lengthening the uptake period, especially for systemic (water soluble) active ingredients which, once they dry on the leaf surface, no longer diffuse rapidly into the leaf. Humectants comprise e.g. hygroscopic fertilizers or salts (urea, AMS), glycerin, glycols, chitosan, petroleum oils and vegetable oils.

Utility Modifiers

Utility modifiers address handling and application issues. **Buffers** can be used to maintain a slightly acid pH in spray solutions, preventing alkaline hydrolysis of susceptible pesticides, and may improve the uptake of certain herbicides.

AMS (ammonium salts) may be added to reduce hard water problems in spray solutions, while antifoam agents suppressed foam that can cause pump and spray problems.

Suspension (dispersing) agents contribute to maintaining particles in suspension, and are particularly relevant for microbial biopesticides as well as conventional pesticides prone to precipitation.

Wetters lower the interfacial tension between the solid particles to ensure that the powder (toxicants) wets and mixes with water in spray tank easily.

Emulsifiers are surfactants with dual lipophilic and hydrophilic characteristics that form solvent-containing micelles to solubilize lipophilic active ingredients (see next chapter).

Additive	Purpose
Solvent	Solvents are used either straight or as mixtures (non-polar solvents + polar co-solvents) to dissolve the active substances.
Dispersing agent	Dispersants stabilize heterogeneous systems such as oil droplets in water or solid particles in water. Used when the system needs to be stable over a long period of time.
Wetter	In the formulation process, a wetter will allow solid particles to be suspended in water.
Emulsifier	Anionic and non-ionic surfactants; lipophilic and hydrophilic surfactant groups form solvent-containing micelles.
Antifoam	To prevent foaming due to the presence of surfactants
Thickener	Thickeners are added to formulations to slow down the sedimentation of oil droplets/solid particles; reduce drift, increase retention (stickers)
Inert carriers / fillers	In solid formulations, used to adjust the ai content to the desired level, prevent caking.
Buffer, water conditioner	Control pH (degradation stability, charge); neutralise hard water
UV filter	To prevent UV degradation
Biocide	Control living contaminants (fungi, algae & bacteria)
Antifreeze	Frost protection

Figure 4.10: Modifier adjuvants affect the physical properties of spray solutions, facilitating handling, safety and application.

Antifoam may be added to counteract the presence of surfactants which frequently leads to foaming in the bottle / container, or in the spray tank.

Thickeners are viscosity modifiers added to reduce drift through increased viscosity or droplet size.

5

FORMULATION TYPES

Agrochemical formulations are the physical mixtures of active ingredients and adjuvants that are used in agriculture and horticulture to control pests, diseases, and weeds.

These formulations can be classified into several types based on the physical state of the active ingredient, the type of adjuvants used, and the method of application.

Each type of formulation has specific advantages and disadvantages, and the choice of formulation depends on the intended use, the target pest or disease, and the environmental conditions.

In this section, we will discuss the various types of agrochemical formulations, their properties, and their uses in agricultural and horticultural applications.

A pesticide formulation is the commercial product as it is sold to farmers, and contains the requisite adjuvants, surfactants, solvents, stabilizers and conserving agents.

The biological activity or mode of action of a pesticide is determined by its active ingredient, which is defined as the ingredient in a pesticide that is biologically active.

The active ingredient in its raw form is called technical material and requires formulation to facilitate its handling, storage and application.

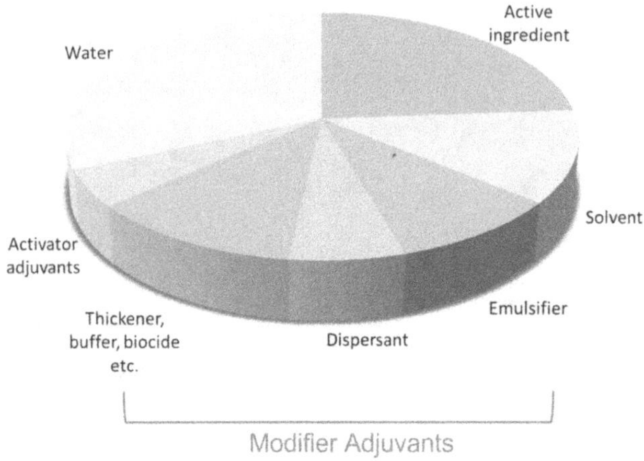

Figure 5.1: Formulation constituents

In addition to the active ingredient, a formulated pesticide may contain water, solvents, modifier adjuvants, dispersants, emulsifiers, with any remaining volume allowing the addition of activator adjuvants such as surfactants to optimize leaf coverage and retention.

Dry (solid) Formulations

Active ingredients may be formulated as so-called dry or solid formulations. These formulations are relatively primitive in that the potential for inclusion of surfactants and additives is limited, and it is thus difficult to optimize their biological effect.

On the other hand, dry formulations can be relatively cheap, stable and light, and are convenient for transport to isolated areas – where they may have to be carried manually – as well as storage under suboptimal conditions.

Solid formulations may be classified based on their constituent active ingredients' aqueous or non-aqueous solubility:

- **Water-soluble active ingredients** solubilize in water, and the spray **solution** is clear, or transparent.
- **Non-soluble active ingredients** are dispersed as particles in **suspension** and will require the addition of dispersing agents and thickeners to prevent their precipitation. On mixing, the spray solution is cloudy.

Attention must be paid to the storage of these formulations, and they may require vigorous shaking prior to use and agitation (stirring) in the spray tank. In general, dispersions of particles sufficiently large for sedimentation are called suspensions, while those of smaller particles are called colloids and solutions.

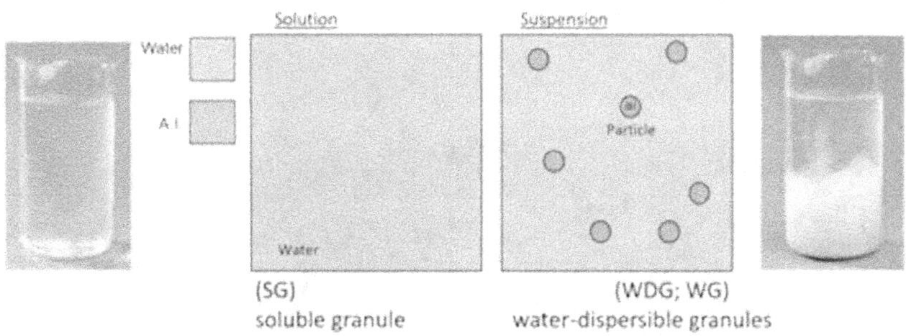

Figure 5.2: Active ingredients may be solubilized (left, clear solution) or suspended (right, cloudy suspension) within the spray solution and in spray droplets.

Water-soluble actives may be formulated as **soluble granules (SG)**; in aqueous spray solution the active ingredient is solubilized, and generally requires the addition of a spreader adjuvant. On mixing, the spray solution is clear, or transparent.

Granules (G) or wettable powders (WP) disperse as suspended particles in the spray tank (cloudy suspension)

They have the advantage of not containing organic solvents but have the disadvantage of low operator safety – an issue which may be addressed by formulating the active ingredients as granules, thereby reducing dust formation and intake while preparing the pesticides for application.

Granules have the advantage of being relatively cheap and are easily packageable, but they are generally only used in or on the soil and may require special equipment for their application.

Water-dispersible granules (WDG) disperse as suspended particles in the spray tank (cloudy suspension). Water-dispersible granules are easily handled (and require minimum packaging) as little dust is produced. However, they require complicated manufacturing procedures (expensive) and their use may lead to nozzle blockage – similar to dusts, wettable powders and other granules.

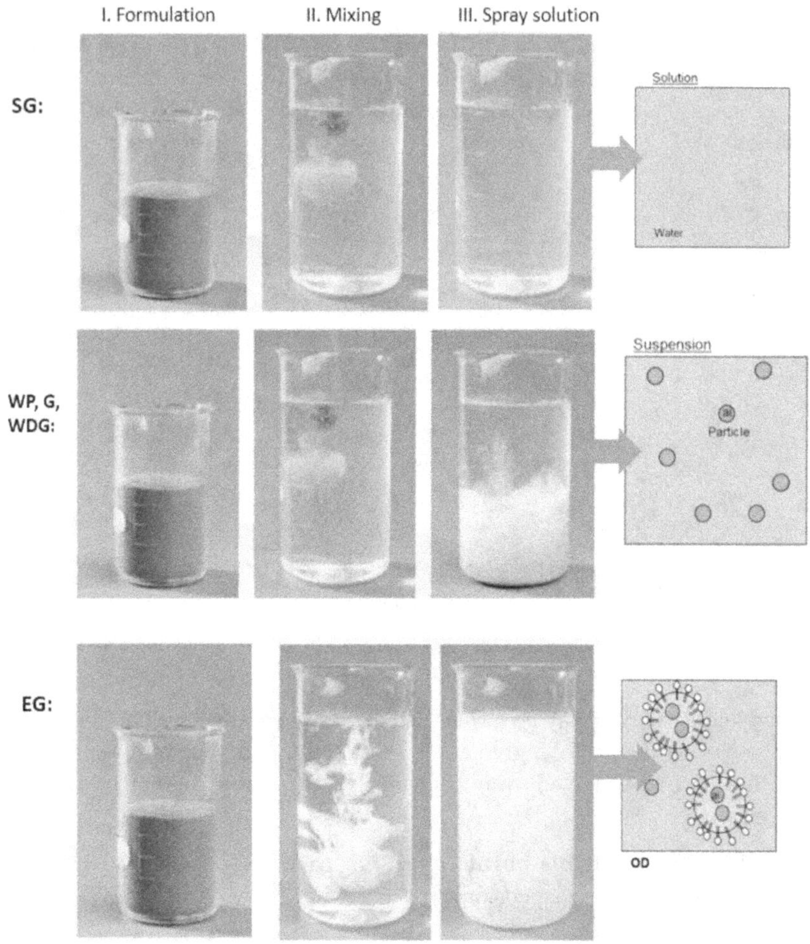

Figure 5.3: Solid formulations as (I) pure formulation, (II) during mixing in water, and (III) their characteristics as spray solutions (transparent) in the spray tank. Photos used with kind permission: Bayer AG, Leverkusen, Germany.

High-loading of the active ingredient without issues of solubility or precipitation is possible, so WDG formulations are suitable for low-volume drone applications.

Tablets are a convenient form of application, required minimum packaging and permit precise dosing, but are expensive and generally only viable for highly active active ingredients.

A **water-emulsifiable granule (EG)** is a formulation consisting of granules to be applied as a conventional O/W emulsion of the active ingredient(s), either solubilized or diluted in an organic solvent, after disintegration and dissolution in water. EG formulations integrate the advantages of both WDG and EC (see under liquid formulations).

Water emulsifiable granules comprise one or several active ingredient(s), either solubilized or suspended in a suitable organic solvent such as oil, and an emulsion is created as the oil (and emulsifiers) create *micelles* containing the solubilized active ingredient.

When emulsion spray droplets dry on the leaf surface, oil droplets remain and may act as adjuvants to facilitate the uptake of lipophilic active ingredients into the lipophilic plant leaf barriers.

Liquid Formulations

Liquid formulations are the most frequently used pesticide products, and permit extensive formulation development through the inclusion of optimized solvents and surfactants, allowing for the optimization of biological activity based on an understanding of plant, pathogen and pest biology.

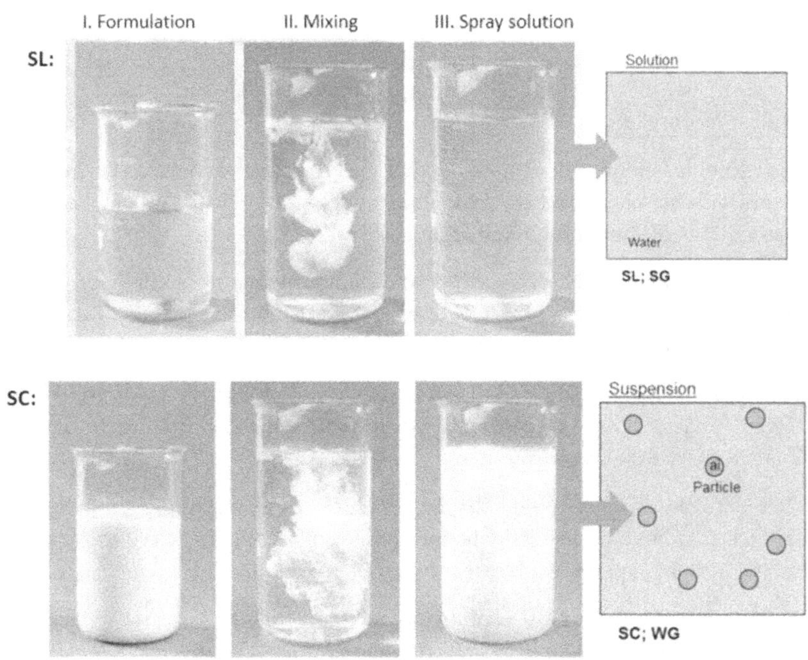

Figure 5.4: SL and SC liquid formulations as (I) pure formulation, (II) during mixing in water, and (III) their characteristics as spray solutions (transparent) in the spray tank. Photos used with kind permission: Bayer AG, Leverkusen, Germany.

Liquid formulations may be classified on the basis of the aqueous or non-aqueous solubility of their constituent active ingredients.

Accordingly, soluble liquid (SL) formulations typically contain water-soluble active ingredients, e.g. Glyphosate, and these products are easy to produce and are stable (precipitates are generally not formed).

In aqueous spray solutions, water-soluble active ingredients formulated as **soluble liquid or soluble granule (SL; SG)** formulations may be solubilized directly, but may require the addition of a spreader.

Alternatively, less soluble active ingredients may be suspended (rather than solubilized) in **suspension concentrate (SC)** formulations, and they disperse as suspended particles in the spray tank (cloudy suspension).

They have the advantage of not containing solvents (thereby reducing their toxicity) while having the disadvantage of poor storage stability (precipitation of the active ingredient).

Dispersant selection is vital when developing these formulations as it strongly influences stability. Attention must be paid to the storage of these formulations, and they may require vigorous shaking prior to use, as well as agitation (stirring) in the spray tank.

Emulsion formulations

Less soluble, more lipophilic active ingredients may be formulated as **emulsifiable concentrates (EC)** which are highly efficacious but have the disadvantage of containing toxic organic solvents (EC).

For EC formulations, the lipophilic active ingredients are solubilized in oil. When mixed in aqueous spray solutions, an emulsion is created as the oil (and emulsifiers) create micelles containing the solubilized active ingredient.

In micelles, the lipophilic tails of the surfactant molecules favor the oil or solvent phase, while the polar hydrophobic "heads" of the surfactant molecules favor the water phase.

When emulsion spray droplets dry on the leaf surface, oil droplets remain and may act as adjuvants to facilitate the uptake of lipophilic active ingredients into the lipophilic plant leaf barriers (wax layers and cuticle).

As these formulations may contain organic solvents, care must be taken to reduce operator exposure. EC formulations may require stirring during application.

SC formulations generally have a lower activity than EC formulations, but may be preferred when a.i.-related phytotoxicity may pose a problem.

Figure 5.5: EC, EW and ME liquid emulsion formulations as (I) pure formulation, (II) during mixing in water, and (III) their characteristics as spray solutions (transparent) in the spray tank. Photos used with kind permission: Bayer AG, Leverkusen, Germany.

If a mix of SC and EC formulations gives surprisingly good results, for example in a research trial, be aware that solvents and emulsifiers from the EC component may increase the uptake of the SC-formulated active ingredient

In contrast to EC formulations, in **emulsion concentrate (EW)** formulations the active ingredient is already dissolved in micelles (a concentrated emulsion in water) and is subsequently diluted in the spray tank. While EC formulations are one of the most used formulation types, EW formulations offer a safer and less environmentally challenging delivery system.

EW formulations have lower solvent contents than EC formulations and are termed "water-based formulations", rather than "solvent-based", and EW formulations offer a safer and less environmentally challenging delivery system. However, reducing the solvent content of emulsion formulations requires complex manufacturing procedures, and this affects the price of the final product.

Microemulsions (ME) are formulations which (together with nanoemulsions) are increasingly drawing attention. A microemulsion has a micelle size below 100 nm (often 10nm) and is water dilutable - microemulsions are thus typically transparent or translucent dispersions of pesticides in aqueous spray solutions.

This small droplet size and the lowered surface tension of microemulsions can give a higher pesticide efficiency and reduce any tendency to coalesce.

Invert Emulsions (EO) contain a water-soluble ai dispersed in an oil carrier. The oil phase also serves as a sticker/spreader to improve rainfastness, leaf coverage and active ingredient uptake.

When applied, invert emulsions form large droplets that do not evaporate or drift easily. Invert emulsions are most commonly used in aerial applications, or in urban environments where drift to susceptible non target plants is a problem. A further advantage is that these emulsion contain the necessary water for the germination of microbial biopesticide spores

<u>Microcapsules and nanoformulations</u>

Microcapsule formulations (CS) require complex manufacturing procedures but permit the solubilization of lipophilic active ingredients in a minimum of solvent, and the protective capsules promote long-term efficacy, as well as reduced toxicity.

A **dispersible concentrate (DC)** is a micronized liquid formulation that is applied as a solid dispersion after dilution in water. A water-miscible carrier is used to dissolve water-insoluble or partially soluble active ingredients. After dilution in water, the pesticidal agents precipitate and disperse as micronized solid particles as the carrier is dissolved in water.

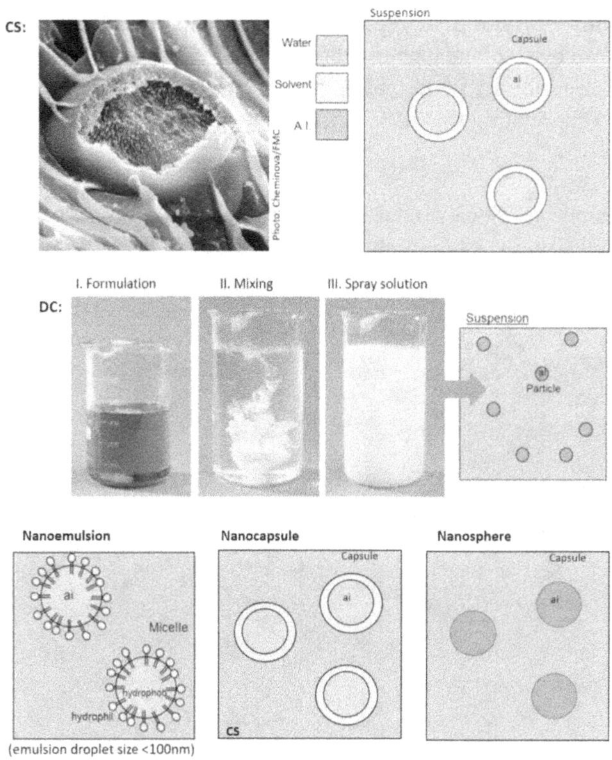

Figure 5.6: Microcapsules (CS), micronized (DC) and nanoformulations.

An advantage of this formulation type is that no milling is required to reduce particle size to ensure optimal efficacy and stability, and is thus suitable for low melting point active ingredients.

The recent application of nanotechnology to agrochemical formulations has created opportunities for developing **nanoemulsions, nanocapsules and nanospheres** that have the potential to improve the efficacy, stability, and effective duration of formulations while reducing their environmental impact.

Formulations for active ingredient combinations

Formulating active ingredients together eliminates the need for tank mixing, which can lead to incompatibility, and facilitates control of a wider range of pests with fewer applications.

Combinations of two or more active ingredients with different physical properties can exist in one formulation. The advantages are that it is possible to formulate multiple active ingredients together, broadening the spectrum of activity and eliminating the disadvantage of tank-mix incompatibility

Oil dispersions (OD) consist of active ingredients particles suspended in oil within micelles and are especially suited for water-sensitive active ingredients, such as sulfonylureas.

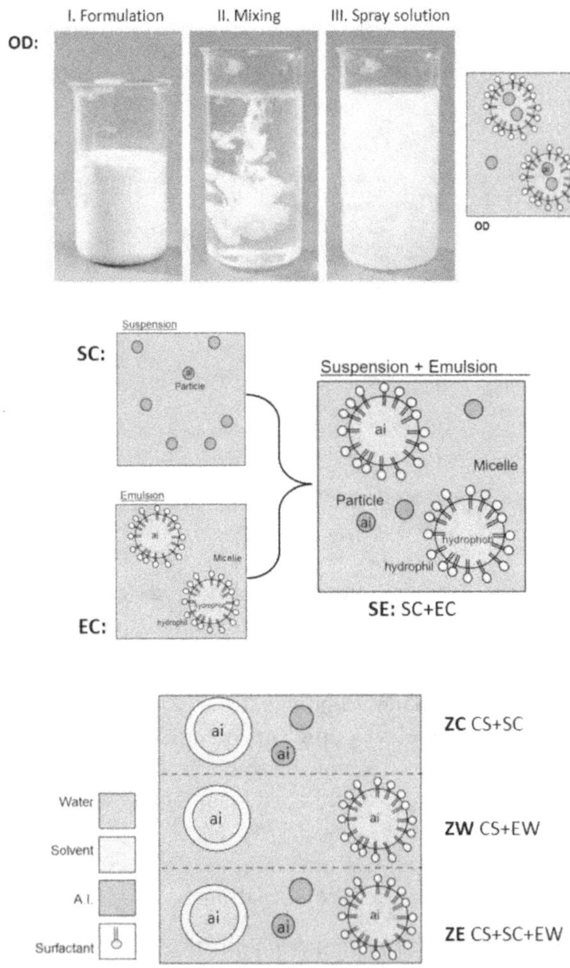

Figure 5.6: Formulations for active ingredient combinations.

The oil phase can have the advantage of acting as an adjuvant to facilitate uptake of the active ingredient into the plant. Attention should be paid to the risk of precipitation (sedimentation) of these formulations – cold storage at low temperatures should be avoided, and the formulations should be shaken before use.

OD formulations may be developed as a replacement for SC formulations. Although SCs are solvent free, the aqueous spray media is normally not ideal to for biological efficacy. Growers add tank mix adjuvants to improve performance. The OD oil phase acts as built in adjuvant, and for water sensitive active ingredients, OD may be the only viable option for liquid formulation.

OD formulations may also be used as an alternative to WDG formulations. WDGs are environmentally safe but are quite expensive, and optimal biological efficacy requires additional adjuvants. The OD with its oil phase can often provide greater efficacy at a lower cost.

Suspoemulsions (SE) provide a combination of suspended particles (SC) and active ingredients solubilized in micelles (EW/EC), and require shaking before use.

Finally, recent formulation types for multiple active ingredients in a single formulation include microcapsule / suspension combinations such as **ZC (CS+SC)**, as well as microcapsule / emulsion (**ZW: CS+EW**) and microcapsule / emulsion / suspension (**ZE: CS+SE**) combinations. Shaking prior to spraying as well as avoidance of operator exposure may be required for these formulations.

Predicting formulation options based on solubility and melting point

Water-soluble actives may be formulated as soluble granules (SG) or soluble liquids; in aqueous spray solution the active ingredient is solubilized.

Solvent-soluble active ingredients may be formulated as emulsions in which the active ingredient is solubilized in the organic phase and the emulsion micelles are suspended in the aqueous spray solution.

For insoluble active ingredients (with low solubility in water or organic solvents) formulation types preferred from a biological point of view such as SL, EC or EW are generally not possible. Here, formulations containing the active ingredient in a solid crystalline form, such as suspension concentrates (SC), ODs or water dispersible granules (WG) are an option, assuming the melting point of the active ingredient is high enough to permit milling to optimal particle sizes.

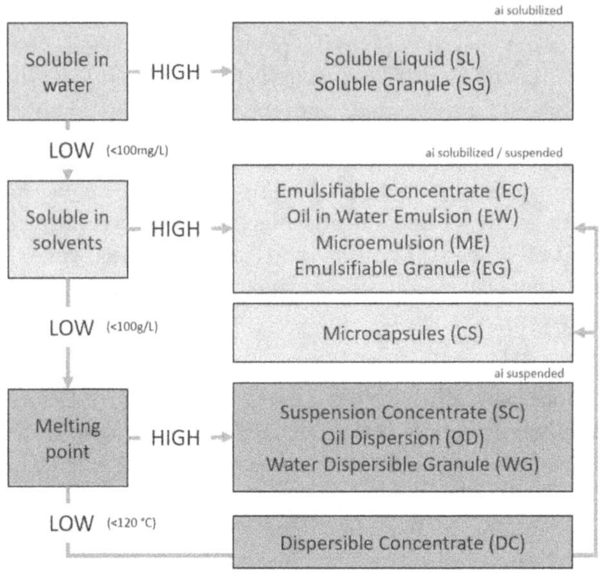

Figure 5.7: Formulation options based on solubility and melting point.

If the melting point is too low to permit milling, dispersible concentrate (DC) formulations can be an option to achieve smaller particle sizes without milling.

Alternatively, formulation emulsions or microencapsulation may still be an option, although this will be technically challenging and product stability may be an issue - this can be expected to increase the final price of the formulated product.

Trends in Formulation and Adjuvant development

Agrochemicals are chemical compounds that are used in agriculture, forestry, and horticulture to protect crops, trees, and plants from pests, diseases, and weeds. They include pesticides, herbicides, and fertilizers, and are typically formulated in a variety of ways, including as liquids, powders, granules, or wettable powders.

Over the past few decades, there have been several trends in agrochemical formulations and adjuvants that have been driven by a variety of factors, including the increasing demand for food, the need to protect against pests and diseases, and concerns about the environmental and human health impacts of traditional chemical agrochemicals.

Formulation trends

A key trend has been the increasing use of biopesticides and biostimulants, which are alternatives to traditional chemical pesticides and fertilizers that are made from natural sources such as plants, microbes, and minerals.

Biopesticides and biostimulants are generally considered to be safer for the environment and human health than traditional chemical pesticides and fertilizers, and are becoming increasingly popular as a way to control pests and diseases and to improve plant growth and health.

	FORMULATION	INERTS, ADJUVANTS
Existing formulations	• SOLID: WP to WG and SC; water soluble packaging - handling, precise dosing • LIQUID: From EC to EW - reduced toxicity (fewer solvents) • Multi-component formulations (e.g. SE; OD; CS) – microplastic-free; multipurpose • Low volume – drone applications, precision agriculture applications • Environmental focus – reduced drift, volatility (herbicides); drought, UV, salinity	• 3rd gen. solid adjuvants; biodegradable polymeric dispersants • Sustainable, CO2 neutral bio-based solvents; "in-can" incorporated adjuvants • Microplastic-free, multipurpose sustainable co-formulants • Adjuvants for precision applications • Adjuvants for adverse environments
New formulations	• Targeted delivery technologies – nanocarriers; natural-origin microencapsulation • Formulations for living (microbial) biopesticides and metabolites	• Adjuvants for targeted and controlled release • Adjuvants for living (microbial) biopesticides and metabolites

Figure 5.8: Trends in agrochemical formulation.

Another trend has been the increasing use of precision agriculture techniques, which use advanced technologies such as sensors, GPS, and drones to more accurately target the application of agrochemicals to specific areas of a field or orchard.

This can help to reduce the amount of agrochemicals that are applied, which can be beneficial both from an environmental and economic standpoint.

There has also been a trend towards the development of more sustainable agrochemical formulations, such as those that are designed to break down more quickly in the environment or that are made from renewable resources.

These formulations can help to reduce the environmental impact of agrochemicals and may also be more cost-effective in the long run.

. . .

Adjuvant trends

A key trend has been the increasing use of adjuvants that are derived from natural sources, such as plant oils and extracts. These adjuvants are generally considered to be safer for the environment and human health than traditional chemical adjuvants, and are becoming increasingly popular as a way to improve the performance of agrochemicals while reducing their environmental impact.

Another trend has been the increasing use of adjuvants that are designed to improve the efficacy of agrochemicals. For example, some adjuvants are designed to improve the sticking or penetration of the agrochemical into the plant, while others are designed to improve the stability of the agrochemical in the environment. These adjuvants can help to improve the effectiveness of the agrochemical, which can be beneficial both from an agricultural and economic standpoint.

There has also been a trend towards the development of more sustainable agrochemical adjuvants, such as those that are designed to break down more quickly in the environment or that are made from renewable resources. These adjuvants can help to reduce the environmental impact of agrochemicals and may also be more cost-effective in the long run.

Overall, the trends in agrochemical formulations and adjuvants have been driven by a combination of factors, including the increasing demand for food, the need to protect against pests and diseases, and concerns about the environmental and human health impacts of traditional chemical agrochemicals. As these trends continue to evolve, it is likely that we will see the development of even more innovative and sustainable agrochemical formulations and adjuvants in the future.

II

HERBICIDE MODE OF ACTION AND FORMULATION

6

HERBICIDE MODE OF ACTION

Herbicides – also known as weed killers – are substances toxic to plants that are used to destroy unwanted vegetation (weeds).

Herbicides control weeds by interfering with their growth. This is achieved through several different Modes of Action (MOA).

Different herbicides work in different ways to kill weeds, and understanding the mode of action can help to determine the best formulation for a given situation.

MOA may be differentiated into Mechanism of Action, referring to the biochemical site of herbicide activity, and Mode of Action – the associated plant symptoms that occur following herbicide application.

In the following, Mode of Action will be used to describe both the site of activity as well as associated plant symptoms.

A sound knowledge of herbicide mode of action is important for determining how to optimally use a herbicide, as well as for understanding the underlying mechanisms of weed resistance and herbicide selectivity.

In herbicide R&D, elucidating herbicide mode of action requires years of research, coupling biochemical and molecular biological knowledge with plant physiology and the timing of symptom development.

Herbicide Classification

Herbicides are classified according to their site, or mode of action. The primary classification based on sites of action includes:

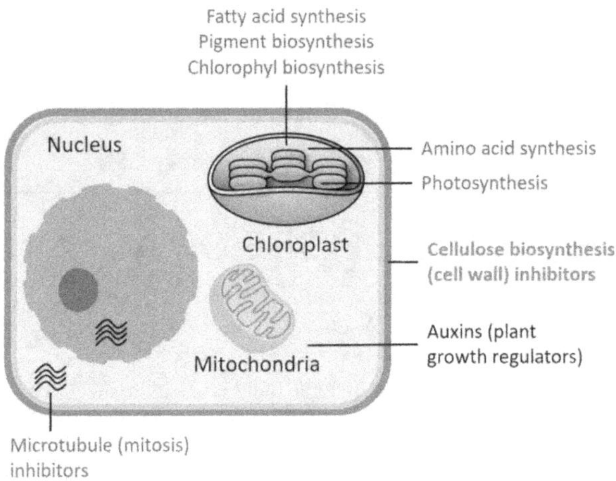

Figure 6.0: Overview of herbicide sites of action.

Photosystem II inhibitors

Photosystem I inhibitors

EPSPS (*5-endopyruvylshikimate-3 phosphate synthase*) inhibitor

HPPD (*hydroxyphenylpyruvate dioxygenase*) inhibitors

PDS (*phytoene desaturase*) inhibitors

ACCase (*acetyl-CoA carboxylase*) inhibitors

PPO (*protoporphyrinogen oxidase*) inhibitors

ALS (*acetolactate synthase*) inhibitors

Glutamine synthetase inhibitors

Synthetic auxins

Cell Division inhibitors

Herbicides and Enzyme Function

Generally, herbicides target specific enzymes and inhibit their normal function.

By inhibiting enzyme function, herbicides prevent the biosynthesis of organic plant constituents, for example fatty acids, which are necessary for plant function.

Fatty acids are an integral component of plant cell membranes and are vital for plant growth and development. Their synthesis in plants is catalyzed by enzymes.

In the following figure, the enzyme functions normally, catalyzing the conversion of precursors to fatty acids.

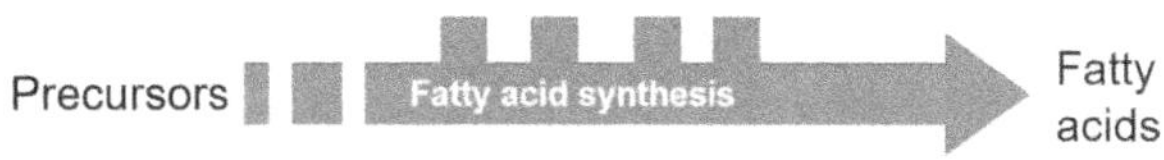

Figure 6.1: Fatty acid biosynthesis.

The herbicide Fenoxaprop inhibits the enzymatic process of fatty acid biosynthesis by specifically binding to the ACCase (acetyl-CoA carboxylase) enzyme.

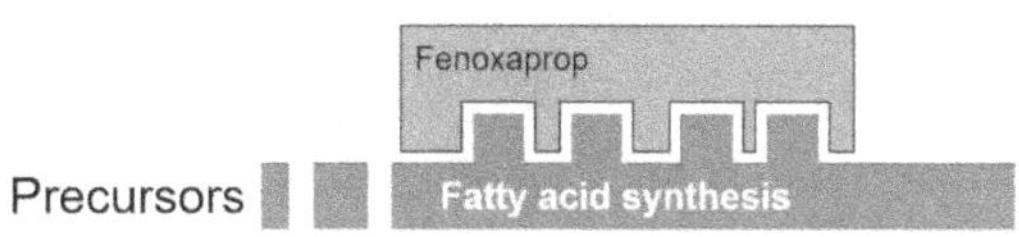

Figure 6.2: Herbicidal inhibition of fatty acid biosynthesis.

This binding is dependent on compatibility between the three-dimensional binding site on the enzyme, and the three-dimensional shape of the herbicide molecule.

Affected plants are unable to synthesize fatty acids and thus new cell and organelle membranes, or maintain and repair existing membranes damaged by reactive oxygen species.

Herbicides may restrict or stop plant growth, give rise to an accumulation of toxic precursor molecules, or they may lead to a decreased scavenging of reactive (radical) oxygen species

Herbicides and Photosynthesis

Most commonly used herbicides affect photosynthesis, either through a direct inhibition of photosynthetic electron transport, through the creation of reactive oxygen species with resultant photo-oxidative damage, or by inhibiting protective mechanisms capable of quenching reactive oxygen species.

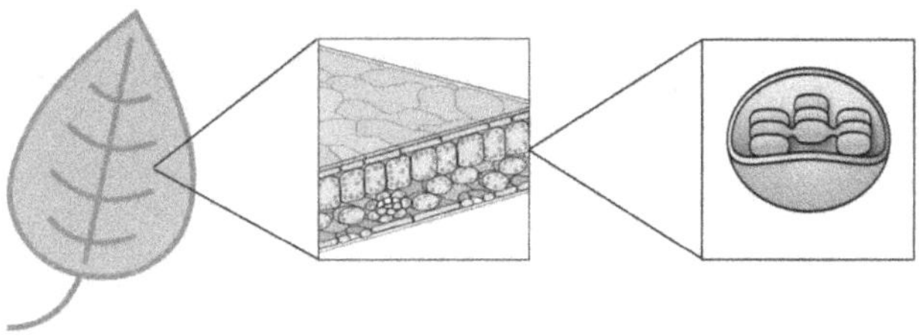

Figure 6.3: The location of chloroplasts within plant leaves.

Photosynthesis is a process which occurs primarily in plants and some bacteria. Thus, herbicides specifically target plants, as opposed to beneficial mammals and (most) microorganisms.

Photosynthesis is the process where plants use light energy to convert CO_2 and H_2O to sugars, with the concomitant release of oxygen.

Photosynthesis primarily takes place in the leaves of green plants. Within plant cells, chloroplasts contain the enzymes necessary for photosynthesis, bound within stacked membrane systems called thylakoids.

From a toxicological point of view, it is important to understand that mitochondrial (respiratory) electron transport is evolutionarily related to photosynthetic electron transport, and that some herbicides may thus affect respiration in mammals, insects and microorganisms.

Photosynthetic Electron Transport

In Figure 6.4, light can be seen to mediate the photolysis of water to oxygen (O), protons (hydrogen ions, H^+) and electrons, and drive these electrons against a redox gradient through PSII and PSI, permitting the reduction of CO_2 to glucose.

Plastoquinone (PQ) mediates the transfer of electrons from PSII to PSI, and contributes to maintaining the redox balance between both photosystems.

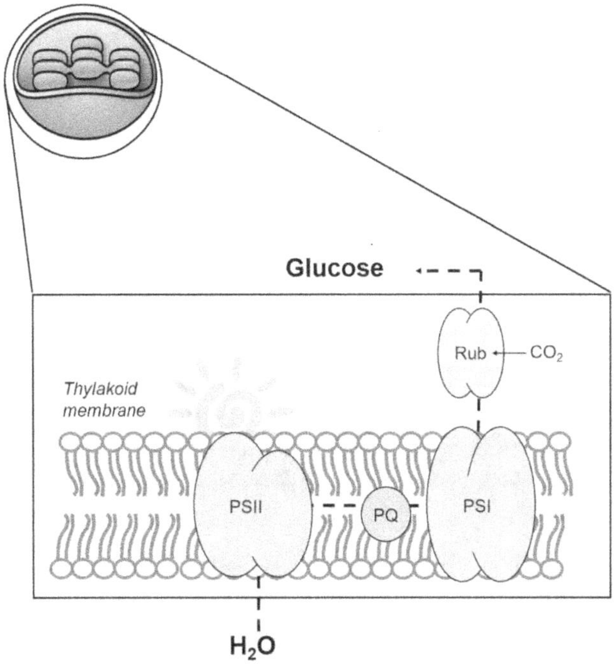

Figure 6.4: The process of photosynthesis in thylakoid membranes.

Herbicidal Photosynthesis Inhibitors

Herbicides of the PSII- and PSI-inhibitor group prevent the normal functioning of electron transport in the photosynthetic apparatus.

This leads to the over-reduction of photosynthetic components and the reaction of excited state electrons with oxygen molecules, leading to the creation of reactive oxygen species (ROS).

Reactive oxygen species (or oxygen radicals) attempt to stabilize their electron states by drawing electrons from surrounding thylakoid membrane, in a destructive process termed membrane peroxidation.

Membrane peroxidation leads to the oxidative destruction of membrane fatty acids – the thylakoid membranes lose their integrity and are no longer able to hold the green chlorophyll molecules which give the leaves their characteristic color.

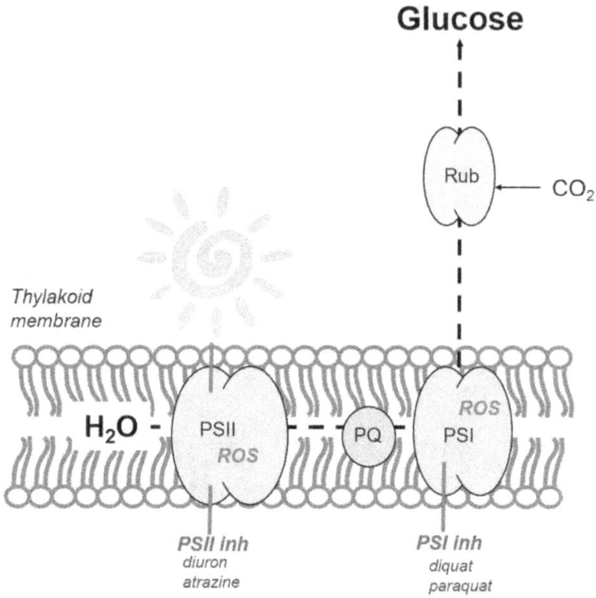

Figure 6.5: The herbicidal activity of PSII- and PSI-inhibiting herbicides.

Accordingly, affected leaves may show typical symptoms of oxidative damage (necrosis, or burning) or bleaching associated with chlorophyll loss.

Inhibition of Aromatic Amino Acid Biosynthesis

Aromatic amino acids are created from carbon building blocks derived from glucose in a process catalyzed by the enzyme EPSP (5-endopyruvylshikimate-3 phosphate synthase) which couples an aromatic carbon ring to precursor molecules (building blocks).

The herbicide glyphosate inhibits the catalytic function of EPSP, thereby inhibiting the formation of aromatic amino acids necessary for plant growth and development.

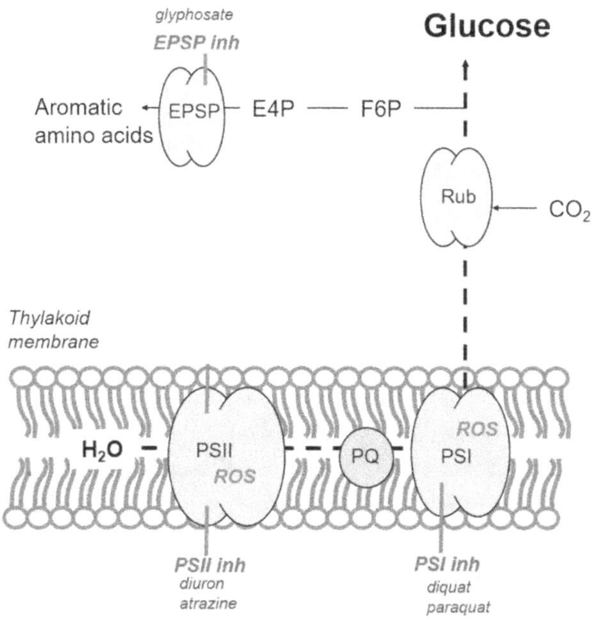

Figure 6.6: Inhibition of aromatic amino acid biosynthesis.

HPPD and PDS Inhibitors

In addition to aromatic amino acid biosynthesis, EPSP catalyses the formation of other aromatic molecules such as the protective antioxidant, Vitamin E – an oxygen radical quencher, and plastoquinone, a mediator of photosynthetic electron transport as well as a precursor in carotenoid biosynthesis.

Vitamin E synthesis is catalyzed by the HPPD (hydroxyphenylpyruvate dioxygenase) enzyme, using aromatic precursors synthesized by the EPSP enzyme.

Several herbicides inhibit the enzymatic activity of HPPD, and these are termed HPPD-inhibitors.

By inhibiting HPPD, these herbicides inhibit the formation of vitamin E, deriving the plant of one of its antioxidant protective mechanisms.

HPPD inhibition also results in a reduced production of plastoquinone. In addition to maintaining the redox balance of the photosynthetic apparatus (thereby minimizing the creation of oxygen radicals due to over-reduction of the photosystems), plastoquinone is a precursor for the production of aromatic carotenoids, catalyzed by the enzyme PDS (phytoene desaturase).

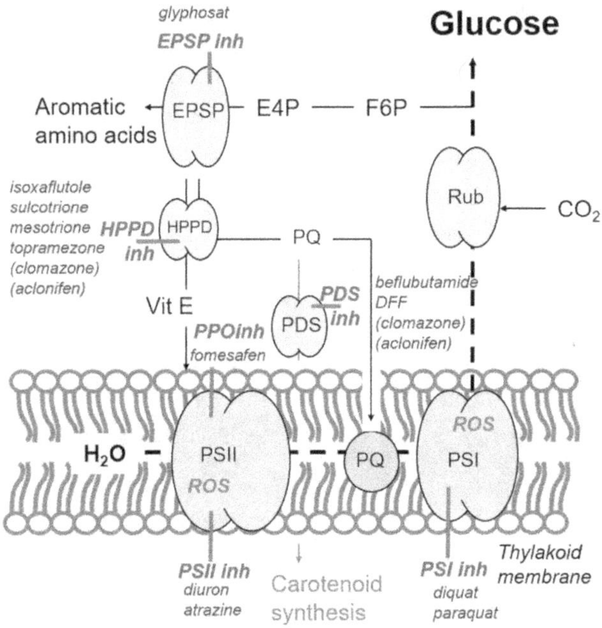

Figure 6.7: The cascade of herbicide inhibition of aromatic molecule biosynthesis in plants.

Like Vitamin E, carotenoids have a protective antioxidant function through their quenching or scavenging of oxygen radicals.

The herbicide Clomazone (previously thought to inhibit the HPPD enzyme) affects plant carotenoid (a reactive oxygen radical quencher) levels, probably through inhibition of DXP (deoxy-D-xylulose 5-phosphate) synthase, a component of carotenoid synthesis in plants.

This herbicidal inhibition of regulatory and antioxidant plant defense mechanisms makes treated plants susceptible to oxidative damage caused by reactive oxygen species arising from the photosynthetic process.

Typical symptoms for HPPD-inhibitory and PDS-inhibitory groups of herbicides include necrotic damage to the leaves due to oxidative damage, and bleaching due to the inhibition of carotenoid biosynthesis, as well as to chlorophyll loss due to reduced thylakoid membrane integrity.

. . .

PPO Inhibitors

Chlorophyll biosynthesis may also be shut down through the herbicidal inhibition of the enzyme PPO (protoporphyrinogen oxidase), which catalyses the formation of chlorophyll from precursors.

This gives rise to the characteristic bleaching of entire leaves by herbicides of the PPO-inhibitor group.

Additional symptoms include oxidative damage (necrosis, burning) due to the accumulation of the photosensitizer protoporphyrin IX in plant cells, leading to the creation of reactive oxygen species and lipid peroxidation.

Inhibition of Branched-Chain Amino Acid Biosynthesis

In addition to the biosynthesis of aromatic amino acids described in the previous section, glucose provides carbon building blocks for all other organic molecules, including branched-chain amino acids.

These amino acids are created from the precursor molecule pyruvate, and this process is catalyzed by the enzyme ALS (acetolactate synthase).

ALS is found in plants and some microorganisms, but not mammals. ALS-inhibitor herbicides prevent the formation of these important amino acids, leading to inhibited plant growth and development.

Inhibition of Fatty Acid Biosynthesis

In addition to its role in branched-chain amino acid biosynthesis, pyruvate is a precursor for the synthesis of fatty acids, catalyzed by the enzyme ACCase (acetyl-CoA carboxylase).

Accordingly, ACCase-inhibitor herbicides prevent the formation of new fatty acid-based plant membranes, including thylakoid membranes.

Furthermore, ACCase-enzyme inhibition results in the creation of reactive oxygen species (ROS), while at the same time inhibiting the plant's ability to maintain and repair membranes damaged by these oxygen radicals.

In addition to stunted growth, susceptible plants treated with this group of herbicides typically show the characteristic red discoloration (caused by phytoalexin-based defense responses) of oxidatively stressed plants.

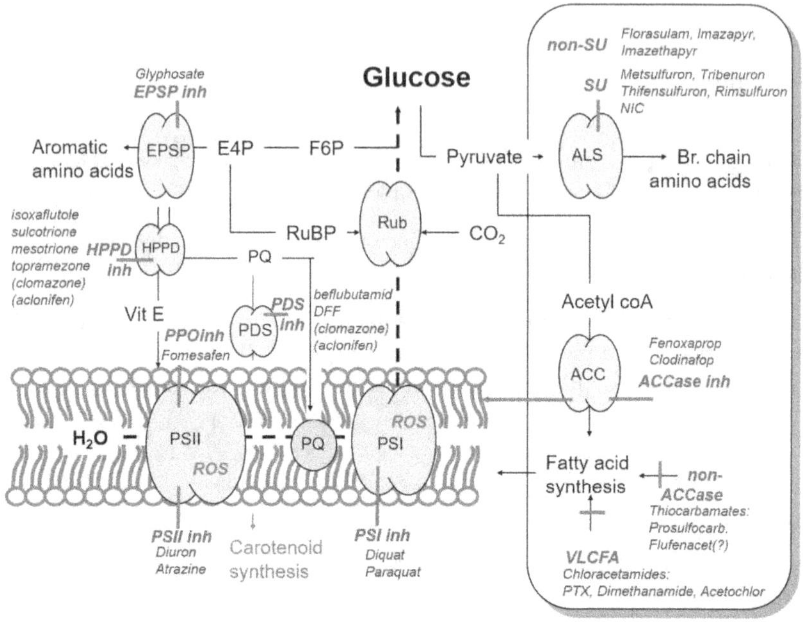

Figure 6.8: Summary of herbicide modes of action related to the photosynthetic apparatus, amino acid and fatty acid biosynthesis.

Other Photosynthetic Herbicide Modes of Action

As described in the previous section, photosynthesis commences with the photolysis of water to oxygen, electrons and protons (hydrogen ions, H^+).

These protons accumulate on the inner thylakoid membrane space (the thylakoid lumen) giving rise to a proton gradient which is used to drive the creation of ATP by the enzyme ATP synthase.

This process of ATP synthesis is termed oxidative phosphorylation. ATP is the universal "molecular unit of currency", and is a biochemical means of storing energy.

Herbicide uncouplers of oxidative phosphorylation collect protons from the thylakoid lumen and transport them through the thylakoid membrane before releasing them.

By lowering the proton gradient, these herbicides reduce the chloroplasts' ability to produce ATP, resulting in the death of the plant.

Glufosinate is a herbicide that inhibits glutamine synthetase, the enzyme catalysing the production of the amino acid glutamine through the metabolism and detoxification of luminal ammonia (NH3). This inhibition leads to reduced protein biosynthesis and elevated ammonia and ammonium (NH4+) levels, resulting in plant death.

Currently, significant research efforts are being undertaken to identify microorganisms that can produce alternative glutamine synthetase inhibitors.

Historically, herbicide development has focused on modes of action related to the photosynthetic electron transport process specific to plants.

However, mitochondrial respiratory electron transport processes are evolutionarily related to photosynthetic electron transport, and herbicides affecting photosynthetic electron transport may adversely affect respiration in mammals, insects and microorganisms.

This has implications for the environmental impact of this group of pesticides, and is the subject of growing research interest.

Alternative modes of action related to photosynthesis – but not to membrane-bound photosynthetic electron transport – might be sought in related, none-thylakoid processes, such as the Calvin cycle – the carbon fixation pathway whereby CO2 is reduced to glucose.

Herbicide Modes of Action Not Related to Photosynthesis

Not all herbicide mode of action is directly or indirectly related to the photosynthetic process, and herbicides may target a number of plant specific metabolic processes within the cell, or within organelles.

Cell wall inhibitors target the biosynthesis of cellulose, an important structural component of the primary cell wall of green plants.

Synthetic auxins mimic native plant auxins, plant hormones which regulate plant growth. At high doses, synthetic auxins are toxic to plants, e.g. through the production of ethylene which can inhibit growth, cause leave drop and ultimately plant death.

Another group of herbicides not related to the photosynthetic process are the microtubule inhibitors, which inhibit cell division and cell elongation in susceptible species.

Although microtubules also play a key role in cell division in animals, microtubule-inhibiting herbicides have a higher specificity for binding plant tubulins than to those of vertebrates.

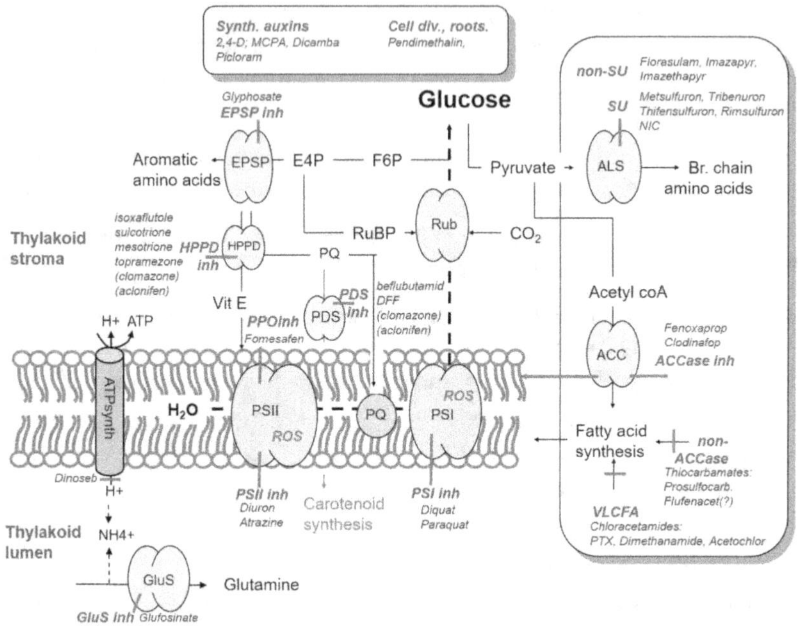

Figure 6.9: Summary of herbicide modes of action. (Abbreviations: PDS phytoene desaturase ALS acetolactate synthase (branched chain: Leu Ile Val) EPSP 5-endopyruvylshikimate-3 phosphate synthase (aromatic Phe Tyr Trp) ACCase acetyl-CoA carboxylase fatty acid synthesis PPO protoporphyrinogen oxidase HPPD hydroxyphenylpyruvate dioxygenase VLCFA very long chain fatty acid ATPsynth ATP synthase GluS Glutamine synthase)

We now have a reference of the key herbicide modes of action and target sites and are ready to move on to an overview of herbicide formulation agrobiology.

7

HERBICIDE FORMULATION AGROBIOLOGY

Herbicides are chemicals that are used to control or kill unwanted plants, also known as weeds. They are widely used in agriculture and horticulture to improve crop yields, reduce the need for manual labor, and prevent the spread of invasive species.

Herbicides can be classified into several types based on their mode of action, chemical structure, and selectivity. Formulating herbicides into effective and stable products is a complex process that involves the selection of appropriate active ingredients, adjuvants, and excipients.

In this chapter, we will discuss the various types of herbicides and their modes of action, and we will explore the factors that are considered when formulating herbicides. We will also discuss the challenges and considerations involved in formulating herbicides for different applications and environments.

An important factor to consider when formulating an herbicide is the solubility of the active ingredient in the herbicide. Herbicides that are highly soluble are generally easier to formulate, as they can be dissolved in a variety of solvents.

However, highly soluble herbicides can also be more prone to leaching, which means they can be washed away by rain or irrigation and may not be as effective in controlling the weeds.

In addition to the active ingredient, herbicide formulations may also contain a variety of other ingredients, including solvents, surfactants, and adjuvants.

Solvents are used to dissolve the active ingredient and make it easier to apply, while surfactants help to improve the spreading and wetting of the herbicide on the plant surface. Adjuvants are substances that improve the performance of the herbicide, such as by helping it stick to the plant or by increasing its efficacy.

Herbicide formulations can be applied to plants in a variety of ways, including as a spray, a dust, or a granule.

The method of application will depend on the specific needs of the crops or plants being protected and the conditions in which they are grown. For example, a spray may be more effective in moist environments, while a dust or granule may be more effective in dry conditions.

It's important to note that herbicides can be harmful to non-target plants, including crops and desirable plants. Therefore, it's important to carefully follow the instructions on the herbicide label and to use the herbicide in the manner recommended by the manufacturer.

Overall, the formulation of an herbicide is an important consideration in its effectiveness and safety.

Careful consideration of the active ingredient, mode of action, solubility, and other ingredients can help to ensure that the herbicide is properly formulated for the specific needs of the crops or plants being protected and the environment in which they are grown.

Herbicide formulation – grass and broadleaf weeds and crops

Herbicide formulation development options are determined by weed biology, and active ingredient physical-chemical characteristics, including:

Target/selectivity – grass or broadleaf weeds and crops

Uptake – foliar, shoots, roots

Application timing – pre or postemergence

Solubility – hydrophilic or lipophilic; solubility in water and organic solvents

Translocation – contact or systemic (xylem, phloem)

Herbicide target/selectivity – grass or broadleaf weeds and crops

Flowering plants (crops and weeds) are divided into two main taxonomic groups based on their morphology, anatomy, and physiology:

Monocotyledons (monocots or grasses) are grass and grass-like flowering crops (rice, wheat, maize) or weeds with one embryonic leaf, or cotyledon.

Dicotyledons (dicots or broadleaves), crops or weeds with two embryonic leaves, or cotyledons.

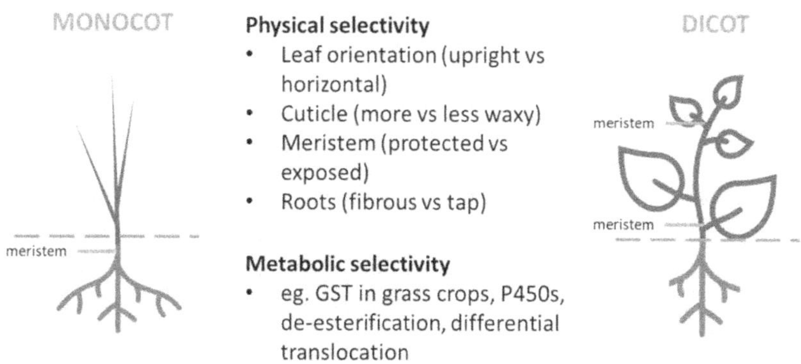

Figure 7.1: Herbicide selectivity – grass or broadleaf weeds and crops

For agrochemical formulation retention and coverage, several morphological differences between grasses and broadleaves are relevant:

Leaf orientation (upright vs horizontal) - due to their upright leaf orientation, grasses can be difficult to reach with herbicide sprays, while the horizontal orientation of broadleaves allows for easier coverage and reduced run-off of the spray droplets.

Cuticle (more vs. less waxy, typically 1–10 μm) - grasses have a thicker cuticular barrier making them more difficult to wet with herbicidal spray solutions.

Stomata (ad- + abaxial) - the lower epidermis (broadleaves) or both lower and upper leaf surface (grasses) contain stomata, respiratory openings which provide an uptake route for pesticide spray solutions.

Meristem (protected vs. exposed) - protected meristems (growing points) at or near the soil surface can make grass growing points challenging to reach with herbicide sprays, while the exposed meristems of broadleaf weeds are more easily reached.

Roots (fibrous vs. tap) - grasses tend to have fibrous root systems which spread near the soil surface, while broadleaves have taproots that penetrate deeply into the soil, below the herbicidal layer (typically the upper 5cm of soil).

The physical or physiological differences between grasses and broadleaves provide opportunities for developing selective formulations that control broadleaf weeds in grass crops or grass weeds in broadleaf crops.

In addition to these examples of physiological selectivity, metabolic selectivity provides additional benefits and opportunities for selective formulation development.

An example of this is the increased glutathione-s-transferase activity found in some cultivated grass species, which, through the use of safeners, allows for the control of grass weeds in grass crops.

Finally, deep-seeding crops below the herbicide layer protects the germinating crop seeds from herbicidal activity, while the greater robustness of cultivated crops permits them to break through the herbicidal layer with little to no damage.

Herbicide application timing - pre- and post-emergence

With respect to application, herbicides may be classified as Pre- or Post-Emergent.

Pre-emergent (PRE-EM) herbicides are applied to the soil, killing the weed seedling when it is germinating. Active ingredient PRE-EM uptake is primarily via the roots and shoots, and soil-applied herbicides do not need additional adjuvants as there is no cuticular barrier in the roots or shoots, neither is there a need for optimal leaf retention and coverage.

However, adjuvants allow water to penetrate soil more easily and/or flow through (infiltrate) the soil, providing benefit when soils have become hydrophobic and will not wet easily. Many pre-emergent herbicides rely on mechanical incorporation or rainfall to activate or reach the root zone.

Crop selectivity to soil-applied herbicides can be a combination of seeding depth (crops are deep-seeded while weed seeds tend to populate the upper soil layer), application timing (e.g., pre-crop application), and differential metabolism in crops.

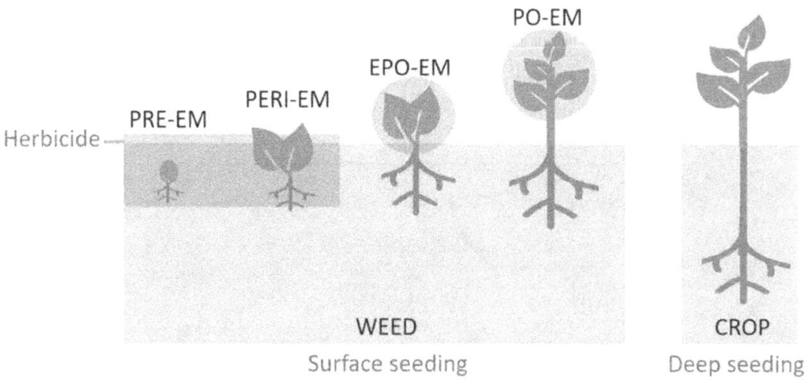

Figure 7.2: Pre- and post-emergent herbicide application.

Post-emergent (PO-EM) herbicides kill weeds after they have emerged from the soil, and may be broken down into peri-emergence application (PERI-EM; when the seedling tips are just visible at the soil surface), early post emergence (EPO-EM; up to the stage when the first two true leaves have developed) and true post-emergent (PO-EM).

Active ingredient foliar uptake requires additional adjuvants to overcome the cuticular barrier and optimal leaf retention and coverage.

One of the factors limiting leaf uptake is the premature drying of droplets on the leaves. The hydrophilic liquid nutrient UAN (urea + ammonium nitrate solution) is known to increase the activity of hydrophilic, foliar-applied herbicides.

Nitrogen-surfactant blends consist of premix combinations of various forms of nitrogen (e.g., ammonium fertilizers) and surfactants and are generally used with herbicides.

The surfactants reduce surface tension and improve leaf surface spreading, while the nitrogen compounds neutralize ionic charges, prevent active ingredient precipitation in the tank mix or on the leaf surface, and promote herbicide uptake primarily into broadleaf weeds.

Crop selectivity to foliar herbicides can be a combination of application timing, differences in physiology between grasses and broadleaf weeds and crops, and differential metabolism in crops ((see previous chapters).

Herbicide acids, esters, and amine salts

Many herbicidal active ingredients are weak acids. Acids are neutral when protonated and negatively charged (ionized) when deprotonated. Bases are positively charged (ionized) when protonated and neutral when deprotonated. Depending on the pH of the soil to which they are applied, herbicides may be neutral, positively charged or negatively charged.

Acid soil	Herbicides	Alkaline soil
+	Cationic	+
Neutral	Weak acid	-
+	Weak base	Neutral

Figure 7.3: Herbicide active ingredient charge in acid and alkaline soils.

A soil's cation exchange capacity (CEC) is a measure of its ability to bind positively charged ions by negatively charged clay or organic matter colloids. The amount of the negative depends on the pH of the surrounding solution.

The pH and CEC of soil determine the solubility and mobility of herbicides, with implications ranging from irreversible binding to soil colloids to leaching to the groundwater.

<table>
<tr><th colspan="2">Cation exchange capacity CEC</th><th>Soil</th><th>Nutrient holding capacity</th><th>Water holding capacity</th><th>Leaching</th></tr>
<tr><td>Low</td><td><10</td><td>Light sand</td><td colspan="2">Low</td><td>High</td></tr>
<tr><td rowspan="2">Medium</td><td rowspan="2">10 to 15</td><td>Dark sand</td><td colspan="2" rowspan="2">Moderate</td><td rowspan="2">Moderate</td></tr>
<tr><td>Light loam</td></tr>
<tr><td rowspan="3">High</td><td rowspan="3">>20</td><td>Dark loam</td><td colspan="2" rowspan="3">High</td><td rowspan="3">Low</td></tr>
<tr><td>Clay Loam</td></tr>
<tr><td>Silt clays</td></tr>
<tr><td>Very high</td><td>>50</td><td>Organic soils</td><td colspan="2">Very High</td><td>Very low</td></tr>
</table>

Figure 7.4: Cation exchange capacity (CEC) and herbicide leaching risk for different soil types

Herbicide acids are often formulated as either amine salts or esters to alter their solubility and activity. Once the amine or ester enters the plant, it is converted to the herbicidally active acid.

Characteristics of herbicide ester and amine salts include:

Esters:

More lipophilic - more soluble when in contact with the cuticle, increased cuticular uptake

More active

More volatile (preferred for lower temperatures)

Amine salts:

More water soluble (higher loading for water-soluble ai's)

Less active

Less volatile (preferred for elevated temperatures)

High electrolyte (salt) herbicide formulations

High-load herbicide salt formulations (such as glyphosate, glufosinate, dicamba), especially when mixed with other salts such as fertilizers are considered "high electrolyte (salt) systems" which pose specific formulation challenges.

Suspension (SC and WG) formulations can experience issues with dispersibility/suspensibility in high electrolyte systems due to reduced surfactant solubility, as the salts reduce the partitioning of many common surfactants between the water phase and active ingredient.

Alternatives to common surfactants include compatibilizers, such as combinations of a hydrotrope (like surfactants but with a much shorter hydrophobic tail) with an amphoteric (zwitterions with dual positive and negative charges) surfactant.

Hydrotropes (sodium alkylbenzene sulfonates, sodium alkylnaphthalene sulfonates, ethanol, urea, and sodium hydroxyalkyl sulfonates) increase the aqueous solubility of slightly lipophilic active ingredients without forming micelles.

Amphoteric surfactants (phosphate esters, betaines, imidazoline derivatives, amino oxides) have dual positive and negative charges and work as compatibilizers to allow different active ingredients to be mixed.

In addition, dispersants may reduce precipitation due to high salt content.

Herbicide Classes and agrobiological groups

Active ingredient physical-chemical characteristics properties may be obtained through various sources, including the Pesticide Properties Database (University of Hertfordshire) and ePesticide Manual (British Crop Protection Council)

It is possible to identify trends and group the key herbicide modes of action based on our understanding of weed biology, herbicide active ingredient physical-chemical characteristics, and target sites.

By differentiating between hydrophilic and lipophilic active ingredients, foliar uptake and uptake through roots and shoots, it is possible to predict the translocation of active ingredients within the plant and understand where the relevant target sites are to be found, identifying barriers for herbicide active ingredient uptake, translocation and target site binding which may be addressed by formulation and adjuvant technologies.

Class	Uptake	Solubility	Translocation	Site
Synthetic auxins	Foliar	hydrophilic	Phloem	Meristem
EPSPS inhibitor	Foliar	hydrophilic	Phloem	Chl. Stroma
Glutamine synth.	Foliar	Hydrophilic	Xylem, phloem	Chl. Stroma
ALS inhibitor	Foliar	Hydrophilic	Xylem, Phloem	Chl. Stroma
HPPD	Foliar	Intermediate, lipophilic	Xylem, phloem	Chl. Stroma
PSII inhibitors	Foliar	Intermediate, lipophilic	Xylem	Thylakoid membrane
ACCase inhibitors	Foliar	Lipophilic	Xylem, phloem	Cytosol
PSI inhibitors	Foliar	Lipophilic	Contact	Thylakoid membrane
DXPS	Roots, shoots	Intermediate	Xylem	Chl. Stroma
PSII inhibitors	Root, shoots	Intermediate, lipophilic	Xylem	Thylakoid membrane
HPPD	Root, shoots	Intermediate, lipophilic	Xylem, phloem	Chl. Stroma
Non-ACCase	Roots, shoots	Lipophilic	(Contact)	Chl. Stroma
VLCFA inhibitors	Roots, shoots	Lipophilic	(Contact)	Chl. Stroma
PPO	Root, shoots	Lipophilic	Contact	Chl. Stroma
PDS inhibitors	Root, shoots	Lipophilic	Contact	Chl. Stroma
Microtubule inh.	Roots, shoots	Lipophilic	Contact	Cytosol, cell wall

Figure 7.5: Herbicide agrobiology groups by class, solubility, uptake, and translocation.

Predictive model for herbicide formulation

Predictive models are valuable tools to help screen initial formulation parameters and focus research on leading prototypes.

The Bromilow model was developed and presented by RH Bromilow et al. in the early 1990s and considers both the partition coefficient (logP) and the charge of the active ingredient molecules. logP and pKa are plotted against each other, and the resulting diagram is a useful model for predicting pesticide mobility in plants at physiological pH.

By placing herbicide active ingredients into the Bromilow model based on their lipophilicity (logP) and partition coefficient (pKa), we can determine and visualize pesticide mobility as well as predict their suitability for pre-emergence (soil) application or post-emergence (foliar) application and uptake.

Using this model, we see that many common herbicides tend towards xylem and phloem mobility; they are systemic and are therefore typically applied as post-emergent (Post em.) foliar treatments, while others are non-systemic and are applied as a pre-emergence (or early post-emergence) soil applications.

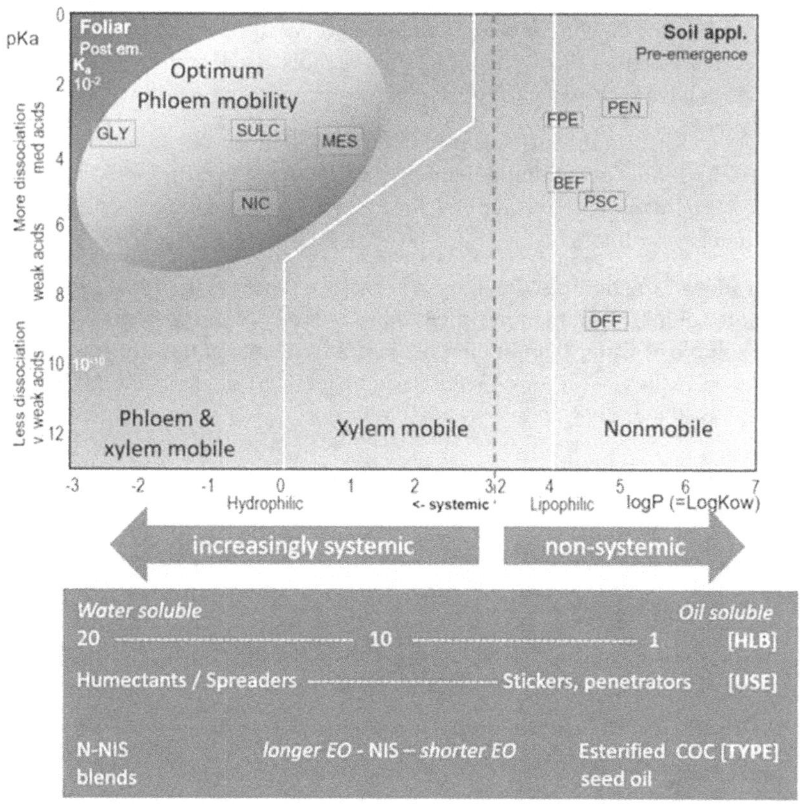

Figure 7.6: Expanded Bromilow diagram for predicting herbicide application and systemicity, as well as formulation types and adjuvants for hydrophilic and lipophilic active ingredients (GLY glyphosate SULC sulcotrione NIC nicosulfuron MES Mesotrione FPE fenoxaprop-p-ethyl BEF beflubutamid DFF Diflufenican PEN pendimethalin PSC prosulfocarb).

Soil-applied herbicides thus tend to be lipophilic and are sprayed on the soil surface, where they function as a waxy barrier to prevent the germination of weed seedlings or are washed into the soil to facilitate root uptake.

Foliar or leaf-applied herbicides pesticides tend to be hydrophilic, and –

depending on the logP and pKa of the active ingredient – translocation of systemic herbicides may take place via the phloem or xylem transport tissues in vascular plants.

The expanded Bromilow model can predict appropriate formulations and adjuvants for specific active ingredients. Because herbicides and surfactants differ in solubility, it is important to match each solubility to maximize their performance as a combination.

Hydrophilic surfactants include oil-in-water emulsifying agents, detergents and solubilizing agents and are suited for foliar-applied hydrophilic active ingredients, such as herbicides. Most surfactants used with post-emergence herbicides have HLB values of twelve or greater.

Lipophilic surfactants include penetrators (oils) and are suited for foliar-applied lipophilic active ingredients. As a rule, however, lipophilic herbicide active ingredients are soil applied for pre-emergence use and do not need additional adjuvants.

Herbicides in the hydrophilic-lipophilic balance (HLB) range from 1 to 10 and are more soluble in oil than in water. Those in the HLB range from 10 to 20 are more soluble in water than in oil. Likewise, surfactants in the HLB range from 1 to 10 work best with an oil-soluble herbicide, and those in the HLB range from 10 to 20 work best with a water-soluble one.

Adjuvant options to optimize pre-emergence root uptake

Surfactants - wetters

Allows water to penetrate soil more easily and/or flow through (infiltrate) the soil. These materials are valuable when soil becomes hydrophobic and will not wet easily.

Humectants

Maintain solubilization of the active ingredient for the plant to absorb. Once out of solution, active ingredients may bind irreversibly to soil colloids. Some humectants may bind water too tightly, not releasing it to the plant, and others may compete with the plant for soil moisture. Organic humectants may be subject to microbial degradation in the soil.

Adjuvant options to optimize post-emergence foliar uptake

Surfactants - wetters

Surfactants with a low HLB are absorbed into the cuticle and increase the fluidity of waxes, increasing the permeance of lipophilic herbicides in the cuticle (crop phytotoxicity). Surfactants with a high HLB (longer EO chains) enhance the hydration of the cuticle and increase the permeance of hydrophilic herbicides. Nitrogen-surfactant blends consist of combinations of nitrogen (e.g., ammonium fertilizers) and surfactants, and are generally used with herbicides. The surfactants reduce surface tension and improve leaf surface spreading, while the nitrogen compounds neutralize ionic charges and prevent the formation of precipitates in the tank mix or on the leaf surface. They are used primarily with broadleaf herbicides.

Humectants

Enhance the hydration of the cuticle and the permeance of hydrophilic herbicides into the cuticle.

Penetrators

COCs enhance spreading and penetration and are used primarily with grass-specific herbicides.

Alkalizers

Basic pH blend adjuvants (blends of non-ionic surfactants, fertilizer, and basic pH enhancer) increase pH to increase charge and water solubility of some herbicides (but can increase selective herbicide phytotoxicity).

Acidifiers

Acidifiers reduce pH to reduce charge (increase lipophilicity) of herbicides to increase cuticular uptake

Water Conditioners

Eliminate or reduce the interaction of ions in the spray solution with the active ingredient. Chelating agents, citric acids, and nitrogen (fertilizer) salts are used as water conditioners.

Herbicide Selectivity, Resistance and Safening

Non-selective herbicides kill all plant material with which they come in contact, while selective herbicides kill weed targets, leaving the crop relatively unharmed.

Herbicides may be selective to plant species possessing a specific enzyme, while weeds may mutate their enzymes and establish a resistant population.

Plants may rapidly metabolize herbicides, or may possess mechanisms to alter the shape and enzyme–binding compatibility of herbicide molecules e.g. by conjugating the herbicide molecule with a glutathione molecule. This is the basis of herbicide safening.

Herbicide Safeners

Herbicide safeners protect crops from herbicide injury by accelerating the metabolism of herbicides, or by inhibiting their translocation within plants, facilitating selectivity between crop plants and weed species targeted by the herbicide.

Figure 7.7: Grass-herbicide treated cereal plants in the presence (left) and absence (right) of a herbicide safener. The safener acts as a catalyst promoting the metabolism of herbicidal compounds in the cereal crop.

Safeners may be applied to crop seeds before planting, or may be applied directly to plants as a mixture with the herbicide.

Herbicide safeners are typically used in monocotyledonous (grass) crops such as cereals, maize, rice and sorghum, and have permitted the development of cereal grass-weed herbicides, in which herbicides are able to target grass weeds without damaging grass (cereal) crops.

Herbicide safening is demonstrated in Figure 7.7, which shows the response of barley seedlings to the application of a selective grass herbicide in the presence (left) and absence (right) of a safener.

Grass herbicides of the aryloxyphenoxy group such as Fenoxaprop and Clodinafop target specific enzymes catalyzing the biosynthesis of fatty acids, thereby inhibiting the synthesis and maintenance of organic plant constituents, such as cell and chloroplast (thylakoid) membranes.

Mechanisms of Herbicide Safening

The aryloxyphenoxy herbicide Fenoxaprop inhibits the enzymatic process of fatty acid biosynthesis by specifically binding to the ACCase (acetyl-CoA carboxylase) enzyme.

This binding is dependent on compatibility between the three-dimensional binding site on the enzyme, and the three-dimensional shape of the herbicide molecule.

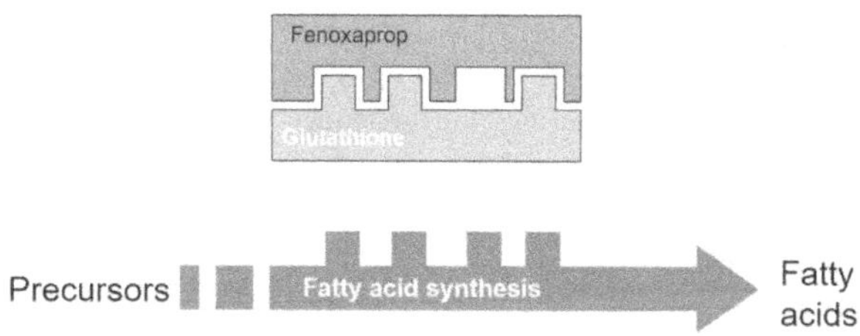

Figure 7.8: Increased glutathione synthesis and glutathione conjugation with Fenoxaprop, following safener induction.

The addition of a safener may increase the synthesis of the naturally occurring glutathione molecule in treated cereals, through its effect on the activity of the enzyme glutathione-S-transferase (GST).

One of the functions of glutathione in plants is to conjugate with foreign compounds (such as xenobiotics) in the first step toward their detoxification and elimination, e.g. through partitioning of the conjugated xenobiotic into vacuoles.

Accordingly, the increased concentration of glutathione in safener-treated cereal crops permits increased levels of glutathione conjugation and detoxification of the herbicide.

Herbicide Safening Is Species-Specific

Natural glutathione levels differ between cereal crop species, and species-dependent differences in safener-induced glutathione concentrations provide differing levels of safening in these crops.

• • •

Herbicide Safening Is Herbicide-Specific

Herbicides safeners are specific to herbicide classes and – as they act on metabolic processes – also to crop species. In cereals, safeners target glutathione conjugation and herbicide metabolism (de-esterification).

The most common grass weed herbicides used in cereals are the Aryloxyphenoxy herbicides Fenoxaprop and Clodinafop, as well as Sulphonylurea (SU) herbicides.

The herbicide safener Cloquintocet increases the activity of GST in cereals, leading to increased glutathione conjugation and detoxification of Fenoxaprop and Clodinafop, while the safener Mefenpyr increases the glutathione conjugation of Fenoxaprop and SUs.

In cereal crops, the safener Fenchlorazole mediates the de-esterification of Fenoxaprop-p-ethyl to its phytotoxic form, Fenoxaprop, as well as its subsequent metabolism to non-toxic metabolites.

In maize and sorghum, safening may be mediated through glutathione conjugation, through the activity of cytochrome P450 enzymes, as well as through translocation inhibition. Importantly, herbicide safeners which are effective in cereals are not effective in maize and sorghum.

For research purposes, it is vital to perform experiments on relevant crops, and not to assume that safeners suitable for one species are applicable to another.

In contrast to safener specificity between cereals and e.g. maize, there is an overlap between safeners appropriate for maize and rice.

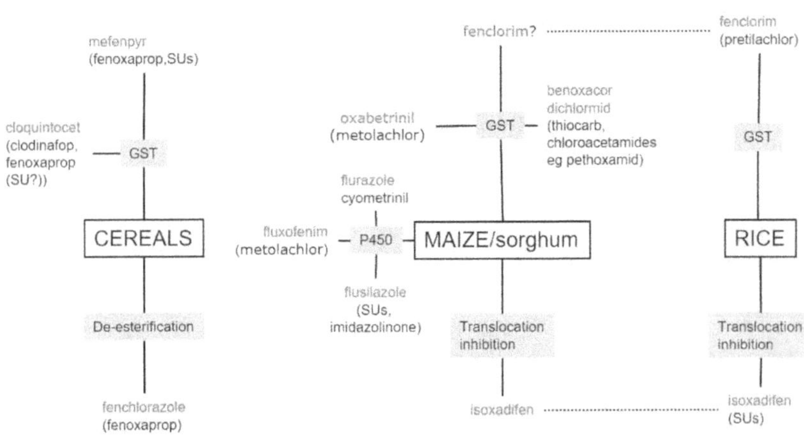

Figure 7.9: Schematic overview of herbicide safeners across species and safener mechanisms.

Thus, herbicide safening in rice may be mediated by the same mechanisms and safeners of glutathione conjugation and translocation inhibition. Notably, however, herbicide safening mediated by P450 metabolism appears to be absent in rice.

Current developments include the identification and development of safeners for use in dicotyledonous crops. For example, there is evidence that insecticides may act as herbicide safeners in soya, and some insecticide safener activity has been observed in crops such as potatoes and oilseed rape.

III

INSECTICIDE MODE OF ACTION AND FORMULATION

8

INSECTICIDE MODE OF ACTION

Insecticides are substances toxic to insects that are used to destroy unwanted pests, as well as their eggs and larvae. Insecticides control pests by interfering with their growth or reproduction, or by repelling insects from treated crops. This is achieved through several different Modes of Action (MOA).

Different insecticides work in different ways to kill insects, and understanding the mode of action can help to determine the best formulation for a given situation.

For example, some insecticides work by disrupting the insect's nervous system, while others work by disrupting the insect's ability to produce energy.

Modes of Action of Insecticides

In the following, Mode of Action will be used to describe both the site of action as well as associated insect symptoms.

The chemical structure of the insecticide active ingredient (AI) usually defines its mode of action by determining its uptake and systemicity, as well as its ability to reach and bind to the target site – the physical location where the insecticide acts.

A sound knowledge of insecticide mode of action is important for determining how to optimally use an insecticide, as well as for understanding the underlying mechanisms of resistance, selectivity and toxicity to beneficial insect species, fish, birds and mammals.

In crop protection R&D, elucidating insecticide mode of action requires years of research, coupling biochemical and molecular biological knowledge with insect physiology and the timing and nature of symptom development.

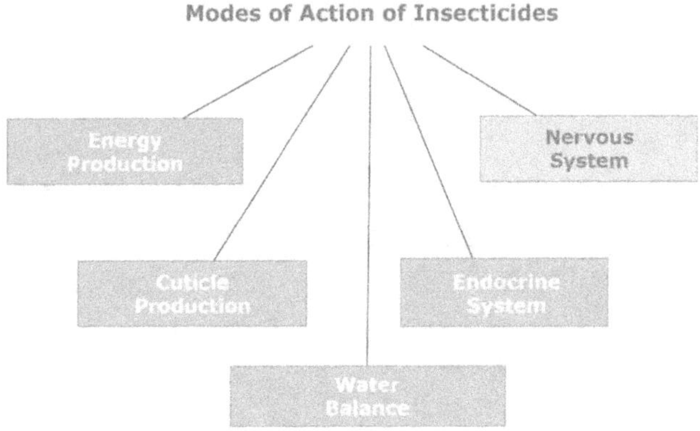

Figure 8.1: Modes of action of insecticides.

Insecticides may target insects through several modes of action. They may target insects by inhibiting energy production, typically by inhibiting mitochondrial respiration, while insecticides such as chitin synthesis inhibitors (CSIs) target the production of insect exoskeletons, and prevent insect molting.

Inhibitors of the endocrine system or insect growth regulators (IGR) act on the hormonal system of insects, preventing them from maturing and reproducing.

Insecticides affecting the water balance of insects (desiccants) break down the waxy coating of their exoskeletons, resulting in water loss and death.

However, most commercial insecticides target the nervous system of insects.

Insecticides Targeting the Insect Nervous System

Most insecticides targeting nervous systems may be grouped into those inhibiting the normal function of Cholinergic and GABA synapses, respectively.

Within these groupings, insecticides target specific neural functions, such as the uptake of ions across neural membranes, by inhibiting specific enzyme functions or by preventing the binding or release of neurotransmitters from their respective receptors.

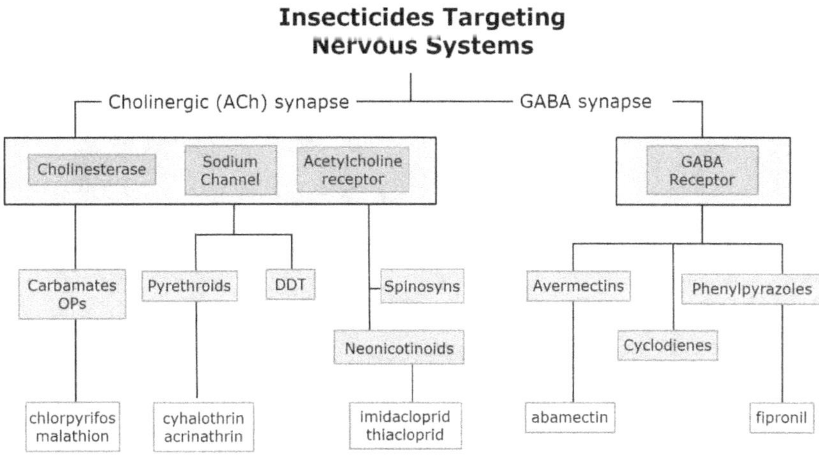

Figure 8.2: Insecticides targeting nervous systems.

Nervous Systems in Insects

To understand the mode of action of insecticides targeting nervous systems, it is necessary to consider how the nervous system functions.

The nervous system of an insect can be divided into a "brain" (comprising three pairs of ganglia, or nerve cell bundles) which are connected to a ventral nerve cord made up of interconnected nerve cells (neurons) along which electrical charges – called impulses – travel.

These impulses are driven by the so-called action potential – a rapidly changing spike of depolarization and repolarization induced by the activation of ion channels in the nerve cell membrane.

It is through this action potential that messages are sent from external sensors to the brain and back.

Neurons may be divided into three categories, depending on their function.

Sensory neurons carry messages from "sense" receptors toward the insects "brain", while motor neurons stimulate responses in muscles and glands. Association neurons transmit signals within the insect's nervous system.

In insects, disruption of the nervous system by insecticides will result in either paralysis or uncontrolled tremors, and ultimately death.

Individual nerve cells consist of two extremities, respectively termed dendrites and axon terminals. Between the axon terminal of the preceding nerve cell (presynaptic neuron) and the dendrites of the succeeding nerve cell (postsynaptic neuron), a junction or synapse is found.

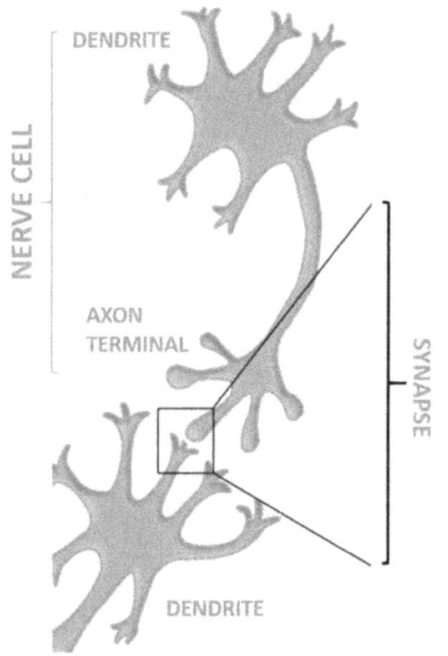

Figure 8.3: Insect nerve cells (neurons) and synapse.

Action potentials are transmitted from the dendrites to the axon terminal, at which stage the impulse must be converted from an electrical potential to a chemical signal capable of traversing the junction.

This chemical signal or molecule is termed a neurotransmitter. Important neurotransmitters include Acetylcholine (ACh) and Gamma Amino Butyric Acid (GABA).

Inhibition of The Cholinergic (ACh) Synapse

The cholinergic synapse uses acetylcholine (ACh) as a neurotransmitter to transfer the action potential from the axon terminal of the presynaptic neuron to the dendrite of the postsynaptic neuron.

These are termed activatory synapses, as they activate the subsequent neuron. In contrast, other synapses such as the GABA synapse are termed inhibitory synapses, as they reduce or inhibit post-synaptic neural activation.

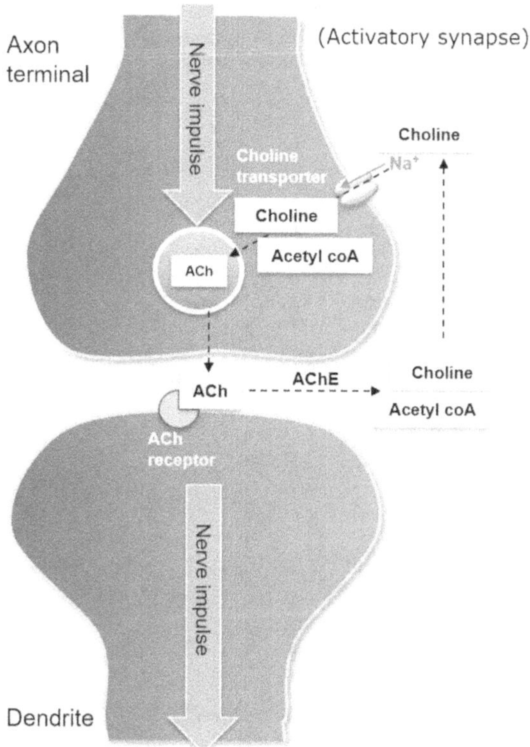

Figure 8.4: Function of the Acetylcholine (ACh) or Cholinergic synapse in insects.

In cholinergic neurons, nerve impulses arriving at the axon terminal trigger the release of ACh from neural vesicles. These neurotransmitter molecules are released from the axon terminal and traverse the synaptic cleft before binding to specific receptors on the surface of postsynaptic dendrites.

Binding of ACh triggers an action potential, and the new nerve impulse is transmitted along the neuron.

The ACh molecule is subsequently released from the receptor to which it was bound (ending nerve stimulation) and is degraded to its constituent components, Choline and Acetyl-CoA, through the actions of the enzyme Acetylcholine Esterase (AChE).

These building blocks are reabsorbed into the presynaptic neuron via voltage-gated sodium (Na^+) channels in the neuronal membrane. Once reabsorbed, these building blocks are used to regenerate ACh neurotransmitters.

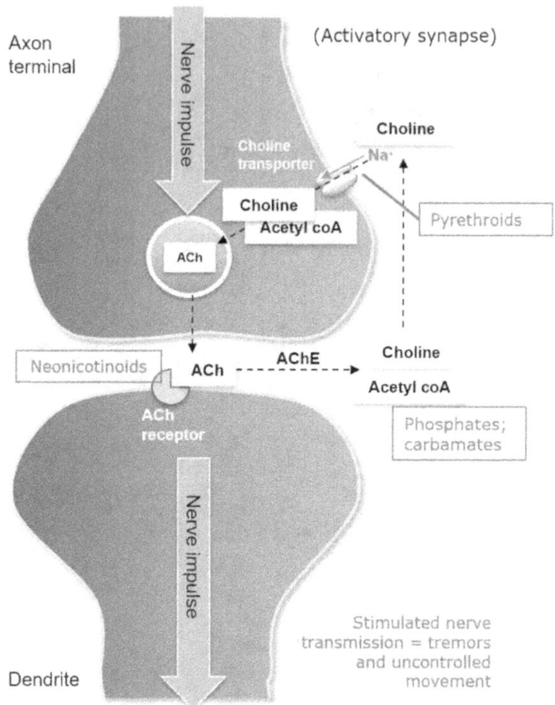

Figure 8.5: Inhibition of the Acetylcholine (ACh) or Cholinergic synapse in insects.

Neonicotinoid insecticides are synthetic analogues of the natural insecticide nicotine. Neonicotinoids compete with ACh neurotransmitters and competitively bind to the ACh receptor on the postsynaptic nerve, resulting in uncontrolled nerve stimulation (hyperstimulation), giving rise to symptoms of hyperactivity and tremors.

Organophosphate and Carbamate insecticides act by inhibiting AChE, preventing it from releasing ACh from its receptor site. In the absence of active AChE, a stimulated nerve cannot return to its resting state, and the result is neural overstimulation and insect death.

Pyrethroid insecticides disrupt normal nerve function by binding to the voltage-gated sodium channels, preventing them from closing normally. This, in turn, gives rise to neural hyperactivity and insect death.

• • •

Inhibition of The GABA Synapse

In GABA (Gamma Amino Butyric Acid) neurons, nerve impulses arriving at the axon terminal trigger the release of inhibitory GABA neurotransmitters from neural vesicles.

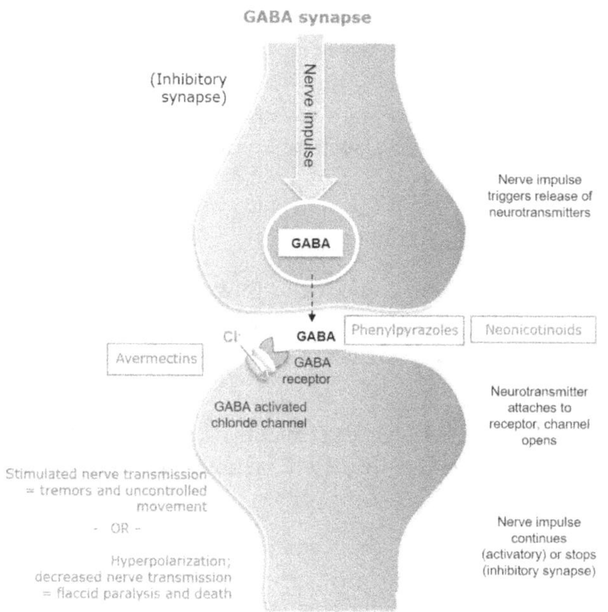

Figure 8.6: Inhibition of the GABA synapse in insects.

These neurotransmitter molecules are released from the axon terminal and traverse the synaptic cleft before binding to specific receptors on the surface of postsynaptic dendrites.

In contrast to activatory Cholinergic (ACh) synapses, GABA neurotransmitters are termed inhibitory neurotransmitters, as they reduce or inhibit post-synaptic neural activation.

One class of GABA inhibitors causes hyperstimulation, by blocking inhibitory GABA neurotransmitters and removing the damping effects of GABA, which maintains normal nerve activity.

Fipronil is a phenylpyrazole insecticide which binds to the GABA receptor, blocking insect-specific GABA-and glutamate-gated chloride channels and causing neural hyperstimulation.

In contrast, Avermectins affect insect-specific glutamate-gated chloride channels in GABA neurons, causing an influx of chloride ions (Cl-) into the postsynaptic neuron – leading to hyperpolarization, paralysis and death.

Neural Agonists and Antagonists

Insecticides targeting nervous systems may behave as agonists (activating the receptors to which they bind, increasing the inhibitory or excitatory effect of neurotransmitters) such as the Neonicotinoids or Avermectins, or antagonists (binding to receptors without activating them and potentially blocking the activity of agonists), such as Fipronil.

Neural Insecticide Specificity

Some insecticides have a high specificity towards insects. For example, it is believed that Fipronil's insect-specificity is due to its greater affinity for insect GABA receptors, compared to GABA receptors in mammals.

Insecticides affecting glutamate-gated chloride channels are specific to insects, as these channels do not exist in mammals. Accordingly, insect-specific insecticides can be considered for use on mammals, for example as flea control products.

In contrast to specific insecticides, other insecticides may target metabolic processes which are common to insects and mammals.

As the AChE enzyme is similar in insects and mammals, this group of insecticides is not considered selective, and appropriate care must be exercised in their application.

Insecticides Not Targeting The Insect Nervous System

Muscular Calcium Channel insecticides include the Diamides e.g., Chlorantraniliprole, which bind and stimulate muscular calcium channels, causing muscle contractions and paralysis.

Insect Growth Regulators (IGRs) include juvenile hormone analogues (e.g. Fenoxycarb) and chitin synthesis inhibitors, such as Lufenuron. IGRs disrupt insect growth, development and reproduction and are generally not effective against adult insects, reducing their impact against beneficial insects, such as pest predators.

Inhibitors of Energy Production target mitochondrial respiration, and include the Pyrrole insecticides e.g., Chlorfenapyr. Other inhibitors such as Sulfuryl

Fluoride and Borax are termed general cellular toxins or non-specific metabolic disruptors (possibly including mitochondrial disruption).

Insect Desiccants include inorganic silica and diatomaceous earth, which disrupt the wax layer of insect exoskeletons, leading to insect dehydration and death.

We now have a reference of the key insecticide modes of action and target sites and are ready to move on to an overview of formulation agrobiology.

9

INSECTICIDE FORMULATION AGROBIOLOGY

Insecticides are chemicals that are used to control or kill insects and other arthropods that are harmful to crops, animals, and humans. They are widely used in agriculture, horticulture, and public health to protect crops from pests, to control vector-borne diseases, and to reduce the risk of damage to structures and other assets.

Insecticides can be classified into several types based on their mode of action, chemical structure, and selectivity. Formulating insecticides into effective and stable products is a complex process that involves the selection of appropriate active ingredients, adjuvants, and excipients.

In this chapter, we will discuss the various types of insecticides and their modes of action, and we will explore the factors that are considered when formulating insecticides. We will also discuss the challenges and considerations involved in formulating insecticides for different applications and environments.

An important factor to consider when formulating an insecticide is the solubility of the active ingredient. Insecticides that are highly soluble are generally easier to formulate, as they can be dissolved in a variety of solvents. However, highly soluble insecticides can also be more prone to leaching, which means they can be washed away by rain or irrigation and may not be as effective in controlling the insects.

In addition to the active ingredient, insecticide formulations may also contain a variety of other ingredients, including solvents, surfactants, and adjuvants. Solvents are used to dissolve the active ingredient and make it easier to apply, while surfactants help to improve the spreading and wetting of the insecticide on the plant surface. Adjuvants are substances that improve the performance of the insecticide, such as by helping it stick to the plant or by increasing its efficacy.

It's important to note that insecticides can be harmful to non-target insects, including beneficial insects such as pollinators.

Overall, the formulation of an insecticide is an important consideration in its effectiveness and safety. Careful consideration of the active ingredient, mode of action, solubility, and other ingredients can help to ensure that the insecticide is properly formulated for the specific needs of the crops or plants being protected and the environment in which they are grown.

Insecticide formulation development options are determined by insect biology, and active ingredient physical-chemical characteristics, including:

Mode of entry - contact vs. ingestion

Solubility – hydrophilic or lipophilic; solubility in water and organic solvents

Translocation – contact, systemic (xylem, phloem), translaminar

---

Insecticide formulation – mode of entry/action

Insecticides are substances toxic to insects that are used to destroy unwanted pests, as well as their eggs and larvae. The primary modes of entry of insecticidal active ingredients into insect pests are via contact and ingestion.

Active ingredient uptake is via direct contact when insects are hit directly with spray droplets. Secondary or indirect contact describes contact with the insecticide because of the insect moving over or resting on sprayed leaves. In both cases, the insecticide is absorbed through the exoskeleton before being transported to the central nervous system.

Contact insecticides are most effective against soft-bodied insects such as Lepidopteran larvae and tend to be less effective against insects with hard exoskeletons or wing covers, as well as insects capable of producing a protective waxy layer (scale bugs) or foam (spit bugs).

Insecticides with stomach action (including those which require conversion to the active form by digestion) are ingested by insects with chewing mouthparts or by sap-feeding insects.

Fumigant insecticides become gases in the air or soil and are used to control nematodes and pathogens in the soil, wood or stored products. Fumigants and insecticides with a vapor phase (such as pyrethroids) enter insects through the tracheal system – this may be the quickest route to the CNS and can lead to a quicker knockdown.

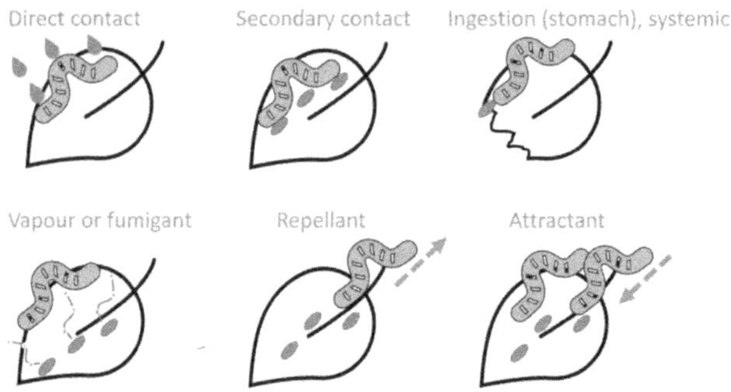

Figure 9.1: Primary modes of entry of insecticidal active ingredients into insect pests are via contact and ingestion.

Repellents are (generally) volatile insecticides or inerts (without insecticidal activity) which prevent insects from feeding on or puncturing treated leaves and fruits.

Repellents used in partnership with contact insecticides can increase insect movement and thus exposure to the contact insecticide and are an important component of insecticide formulation development. Insecticides with antifeedant modes of action provide similar benefits for protecting high-value fruits and vegetables.

Attractants include pheromones which affect insect communication, preventing males from locating females. Formulation inerts, especially sugar-based inerts, can attract insect pests away from the crop and toward sacrificial plants.

Insecticide uptake and action – contact and systemic

With respect to application, insecticides may be classified as contact, translaminar or systemic.

Contact insecticides remain on the leaf surface and target existing populations (cuticular uptake) or new infestations. Contact insecticides tend to be more lipophilic (less systemic) than systemic insecticides, thus new growth is not protected.

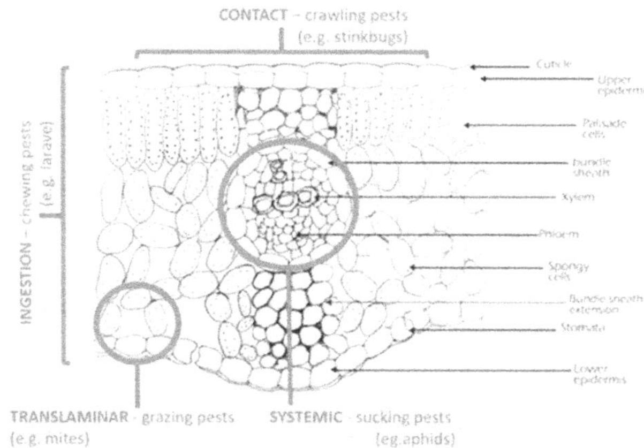

Figure 9.2: Insect targets and leaf distribution of contact, translaminar and systemic insecticides

Locally systemic (translaminar) insecticides penetrate leaf tissues, translocate to the leaf underside, and form a reservoir of active to provide residual activity against certain foliar-feeding insects and mites.

Systemic **insecticides** are absorbed by and transported through plants, target feeding pests and new growth. Systemic insecticides tend to be more hydrophilic than contact insecticides.

The primary benefits of **contact insecticides** are that they target chewing as well as non-feeding insects, have a rapid knockdown (their lipophilic characteristics ensure rapid uptake across the insect exoskeleton or integument, and many contact insecticide formulations have repellent activity (often due to the organic solvents used to solubilize the lipophilic active ingredients).

In addition, they have fewer residues (remain within regulatory Minimum Residue Levels, MRLs) as they rarely penetrate plants. Finally, as contact insecticides tend to be based on older chemistries, they have multisite modes of action (rather than the more specific modes of action of more modern, systemic insecticides), providing them with resistance management benefits.

Many contact insecticides are not suitable for use at bloom time due to negative effects on non-target flying beneficials (bees). Their disadvantages also include shorter residual duration (due to environmental exposure and degradation), and thus they require more frequent reapplication.

Finally, by remaining on the leaf surface, they are subject to wash-off by rain, leading to reduced efficacy and soil contamination.

Formulation development options include increasing rainfastness and UV stability, and the development of biodegradable repellent solvents (EC formulations)

Systemic insecticides can protect new growth and target sucking pests and pollen feeders. As they penetrate the plant, they tend to be more rainfast and UV stable, providing long residual duration and requiring fewer applications.

Their systemicity does, however, mean they can reach pollen and flowers and may thus negatively affect beneficial insects. Systemicity is also associated with higher residues (MRLs) and the risk of phytotoxicity through systemic accumulation.

Systemic insecticides are relatively modern, more selective chemistries, and they tend to have a single-site mode of action. To address resistance issues, they are typically used in combinations with other insecticidal active ingredients, which can introduce compatibility issues that must be addressed by appropriate formulation and adjuvant technologies.

Formulation development options include optimizing uptake kinetics to improve efficacy while avoiding phytotoxicity, and optimizing formulations for compatibility in tank mixes.

Translaminar insecticides are able to pass through the leaf from the upper (sprayed) to the lower (unsprayed) surface. Due to their translaminar activity, they can target grazing insects such as mites and thrips, which hide on the underside of leaves.

Due to their (locally) systemic uptake into leaves, they tend to be more rainfast and UV stable, proving long residual duration and requiring fewer applications. As for systemics, this uptake is also associated with higher residues (MRLs), although the risk of phytotoxicity through systemic accumulation can be lower.

Formulation development options include optimizing uptake kinetics to improve efficacy while avoiding phytotoxicity and optimizing formulations for compatibility in tank mixes.

Insecticide Classes and agrobiological groups

Active ingredient physical-chemical characteristics properties may be obtained through various sources, including the Pesticide Properties Database (University of Hertfordshire) and ePesticide Manual (British Crop Protection Council)

It is possible to identify trends and group the key Insecticide modes of action based on our understanding of insect biology, Insecticide active ingredient physical-chemical characteristics, and insecticidal activity.

Family	AI	logP	Systemicity	Activity
Pyridine	Pymetrozine	-0,2	Systemic, translaminar	Ingestion (sucking)
Tetronic acid	Spirotetramat	2,5	Systemic, translaminar	Ingestion & contact
Neo- nicotinoids	Imidacloprid	0,6	Systemic, translaminar	Ingestion (sucking)
Carbamates	Carbofuran	1,5	Limited systemic, translaminar	Contact & ingestion
Diamides	CTPR	2,8	Limited systemic, translaminar	Ingestion (& contact)
IGRs	Diflubenzuron	3,9	Contact, translaminar	Contact & ingestion
Phenyl-pyrazole	Fipronil	4	Contact	Contact & ingestion
Avermectins	Abamectin	4,4	Contact, translaminar	Contact & ingestion
Spinosyns	Spinosad	4,5	Contact, translaminar	Contact & ingestion
Organo-phosphates	Chlorpyrifos	4,7	Contact	Contact & ingestion
Pyrethroids	Cyhalothrin	6,9	Contact	Contact & ingestion

Figure 9.3: Insecticide agrobiology groups by class, solubility, uptake, and translocation.

By differentiating between hydrophilic and lipophilic active ingredients and contact and systemic activity, it is possible to predict barriers for Insecticide active ingredient uptake, systemicity, and activity which may be addressed by formulation and adjuvant technologies.

Predictive model for insecticide formulation

By placing Insecticide active ingredients into the predictive Bromilow model based on their lipophilicity (logP) and partition coefficient (pKa), we can determine and visualize pesticide mobility and predict their suitability for contact and systemic activity.

Insecticides may be hydrophilic and show great systemicity within the plant, making them ideal for combating sucking pests such as aphids, or lipophilic and non-mobile, making them suitable for controlling rasping, surface-feeding pests such as mites and thrips, as well as chewing insects.

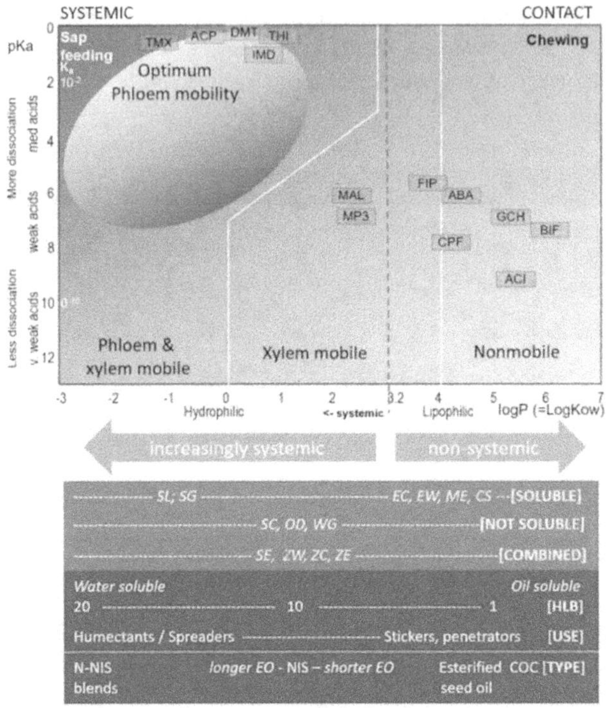

Figure 9.4: Expanded Bromilow diagram for predicting insecticide application and systemicity, as well as formulation types and adjuvants for hydrophilic and lipophilic active ingredients (TMX thiamethoxam ACP acetamiprid DMT dimethoate IMD imidacloprid THI thiacloprid MAL malathion MP3 methyl parathion FIP fipronil ABA abamectin CPF chlorpyrifos GCH gamma-cyhalothrin ACI acrinathrin BIF bifenthrin)

The expanded Bromilow model can be used to predict appropriate formulations and adjuvants for specific active ingredients. Because insecticides and surfactants differ in solubility, it is essential to match the solubility of each to maximize their performance as a combination.

For hydrophilic active ingredients, solubilization in the aqueous spray solution in the form of soluble liquids SL or soluble granule SG formulations is generally possible, while for lipophilic active ingredients that are soluble in organic solvents such as oil, emulsion formulations or microcapsule formulations are appropriate. If multiple active ingredients are to be mixed, SE, ZW, ZC or ZE can be considered.

• • •

Formulation options for systemic insecticides

Surfactants - wetters - Surfactants allow spray droplets to spread beyond their initial contact area, providing a greater area of uptake for systemic active ingredients, or a more homogenous distribution on the leaf surface for active ingredients with contact activity.

Surfactants with a low HLB (shorter EO chains) are absorbed into the cuticle and increase the fluidity of waxes, increasing the permeance of lipophilic insecticides in the cuticle (possible crop phytotoxicity).

Crop oils and COCs - Crop oils and crop oil concentrates COCs spreading and penetration of some (high MW, more lipophilic) systemic and translaminar insecticides. Can induce phytotoxicity.

Humectants - Enhance the hydration of the cuticle and the permeance of hydrophilic insecticides into the cuticle.

Formulation - Emulsions tend to be more effective than suspensions; suspension formulations often allow higher loading and are less phytotoxic.

Hydrophilic surfactants include oil-in-water emulsifying agents, detergents, and solubilizing agents and are suited for foliar-applied hydrophilic insecticides.

Lipophilic surfactants include penetrators (oils) and are suited for foliar-applied lipophilic active ingredients. As a rule, however, lipophilic insecticides that are soil applied against nematodes do not need additional adjuvants as there is no cuticular barrier in roots.

Formulation options for contact insecticides

Surfactants - wetters

Surfactants allowing spray droplets to spread beyond their initial contact area, providing a more homogenous distribution on the leaf surface for active ingredients with contact activity.

Stickers

Stickers increase droplet or particle adherence to the leaf surface and prevent run-off on application as well as wash-off during heavy rainfall. Spreader-stickers contain surfactants (spreading) and stickers (increase adhesion)

Crop oils and COCs

COCs enhance spreading of contact insecticides and can improve rainfastness but can lead to leaf penetration and phytotoxicity.

Formulation

Suspension concentrate formulations are generally retained better than wettable powder formulations.

IV

FUNGICIDE MODE OF ACTION AND FORMULATION

10

FUNGICIDE MODE OF ACTION

Fungicides are synthetic or natural chemical compounds or biological organisms able to kill or inhibit fungi or fungal spore germination, as well as Oomycetes. This is achieved through several different Modes of Action (MOA).

The chemical structure of the fungicide active ingredient (AI) often defines its mode of action by determining its uptake and systemicity, and its ability to reach and bind with its target site – the physical location where the fungicide acts.

A sound knowledge of fungal biology and fungicidal mode of action is important for determining how to optimally use a fungicide, and for understanding the underlying mechanisms of resistance and selectivity.

Different fungicides work in different ways to kill fungi, and understanding the mode of action can help to determine the best formulation for a given situation.

For example, some fungicides work by disrupting the fungi's cell membranes, while others work by inhibiting the fungi's ability to produce energy.

Mode of Action

In the following, fungicides and their modes of action are classified into three broad groups:

Inhibitors of ergosterol synthesis.

Inhibitors of mitochondrial electron transport (respiration inhibitors).

Multi-site enzyme inhibitors, nucleic acid and protein synthesis inhibitors.

Inhibitors of Ergosterol Synthesis

Because ergosterol is a component of fungal cell membranes, but absent in animal and plant cell membranes, its synthesis is an important target for fungicides.

Ergosterol is formed from its precursor squalene via a multistep ergosterol synthesis pathway, and is necessary for membrane structure, integrity and function.

Demethylation inhibitors (DMIs) are a group of systemic fungicides targeting cell membrane integrity through the inhibition of C14-demethylation during ergosterol biosynthesis.

DMIs include Imidazoles, Pyrimidines and Triazoles with primarily curative activity.

This group of fungicides has curative systemic activity and – in contrast to DMIs – is effective against Ascomycetes. By having multiple sites of action, the risk of resistance development is low for this group.

Morpholines can be used in tank or formulated mixtures with DMIs, broadening the spectrum of activity and reducing the risk of resistance.

Morpholines tend to be less costly than DMIs, and combinations of these products can reduce the overall cost of pathogen control.

Although both Dimethomorph and Fenpropimorph are chemically classified as morpholines, the former is an inhibitor of cellulose synthase and acts against Oomycetes whereas the latter is an inhibitor of ergosterol biosynthesis and acts against foliar Ascomycetes.

To target oomycetes, dimethomorph may be formulated differently to fenpropimorph, for example through the addition of super spreaders to promote stomatal uptake.

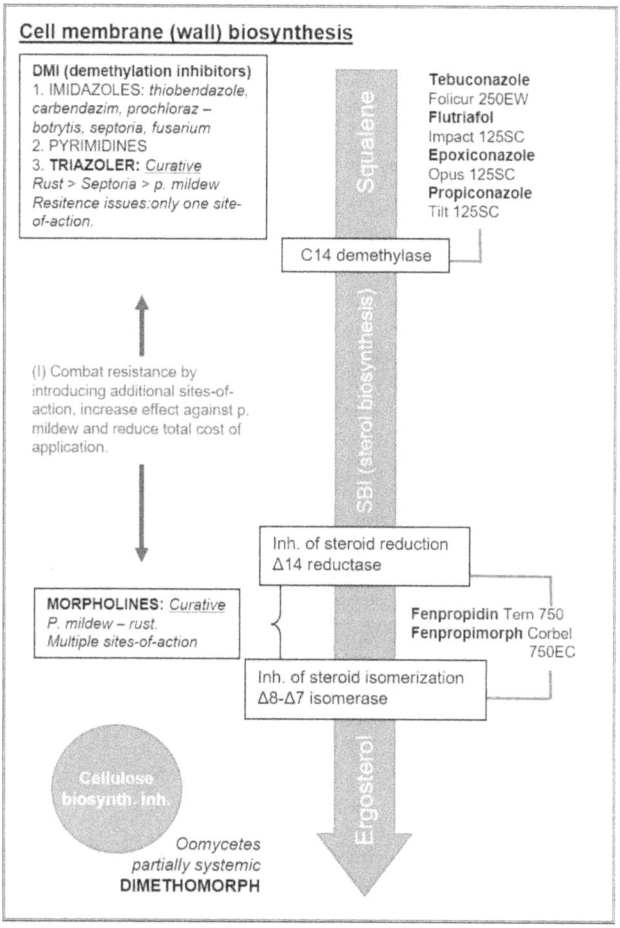

Figure 10.1: Summary of key inhibitors of Ergosterol Synthesis and their sites of action.

Inhibitors of Mitochondrial Respiration

The second major group of fungicides are inhibitors of mitochondrial respiration.

Strobilurins form part of the group of *quinone-outside inhibitor (QoI) fungicides,* inhibiting mitochondrial respiration at the Qo site of cytochrome b, part of the cytochrome bc1 complex (Complex III), preventing spore germination and mycelial growth in fungal pathogens.

Strobilurin fungicides can be both partially systemic and protectant in action. They are active against most cereal fungal diseases, including Septoria, rusts and mildew.

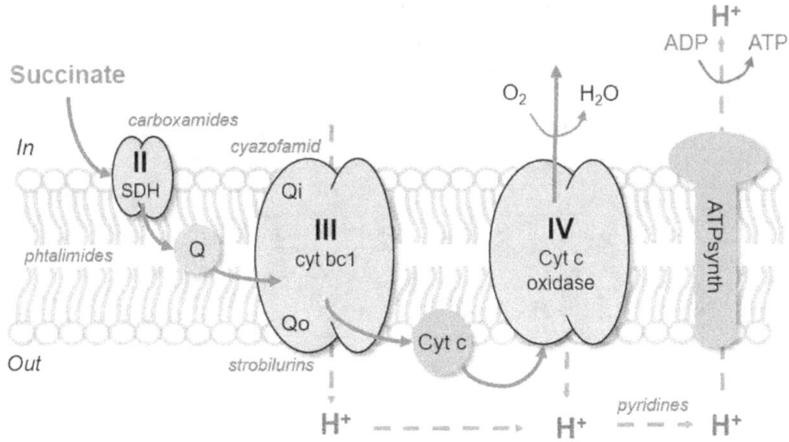

Figure 10.2: Modes of action of mitochondrial respiration inhibitors.

As mitochondrial respiration also occurs in Oomycetes, this group of fungicides may also control pathogens such as *Plasmopara viticola*, responsible for downy mildew.

Strobilurins also have physiological benefits, maintaining green leaf area for longer (termed 'greening') and delaying crop senescence.

Strobilurins are generally as protectively effective as triazoles, although triazoles are more systemic and thus have greater curative activity.

Strobilurins have a great residual long-term activity, and can be used in tank or formulated mixtures with triazoles, broadening the spectrum of activity and permitting greater flexibility of spray timing

Also inhibiting mitochondrial respiration at Complex III, Cyazofamid is an Oomycetes-selective *quinone-inside inhibitor (QiI) fungicide.*

Carboxamides form part of the group of *succinate dehydrogenase inhibitors (SDHIs),* inhibiting the activity of mitochondrial Complex II (succinate dehydrogenase) and thus respiration in fungal cells

As Complex II is a common mitochondrial enzyme system found in eukaryotic and prokaryotic organisms, non-target effects of complex II inhibitors must be considered.

The Pyridine group of fungicides are *proton gradient uncouplers,* inhibiting mitochondrial respiration in fungal pathogens by uncoupling the proton gradient and reducing the proton driven synthesis of ATP via the enzyme ATP synthase.

Folpet and the phthalimide fungicide Captan are *thiol reactants,* blocking the activity of thiol-containing enzymes involved in mitochondrial respiration in fungal spores.

Their protective action is mainly due to inhibition of spore germination. As for many other respiration inhibitors, they are also active against certain Oomycetes, such as Pythium.

Other Fungicide Modes of Action.

Multi-site enzyme inhibitors such as Chlorothalonil and Thiram inhibit several thiol-containing enzymes involved in respiration, while *single site fungicides,* such as Epoxiconazole and Azoxystrobin only inhibit a single, specific metabolic process.

As pathogens struggle to overcome multiple metabolic processes, the risk of resistance is low for multi-site fungicides. Besides multi-site fungicides, risk management programs are based on the use of multiple active ingredients with different modes of action.

Nucleic acid metabolism and protein synthesis inhibitors include benzimidazoles, phenylamides and dicarboximides, and inhibit DNA as well as RNA synthesis, affecting cell division and cellular metabolism.

We now have a reference of the key Fungicide modes of action and target sites and are ready to move on to an overview of fungicide formulation agrobiology.

11

FUNGICIDE FORMULATION AGROBIOLOGY

Fungicides are chemicals that are used to control or prevent fungal diseases in crops, plants, and animals. They are widely used in agriculture and horticulture to protect crops from fungal diseases, to improve crop yields, and to maintain the quality and appearance of plants.

Fungicides can be classified into several types based on their mode of action, chemical structure, and selectivity.

Formulating fungicides into effective and stable products is a complex process that involves the selection of appropriate active ingredients, adjuvants, and excipients.

In this chapter, we will discuss the various types of fungicides and their modes of action, and we will explore the factors that are considered when formulating fungicides. We will also discuss the challenges and considerations involved in formulating fungicides for different applications and environments.

An important factor to consider when formulating a fungicide is the solubility of the active ingredient in the fungicide. Fungicides that are highly soluble are generally easier to formulate, as they can be dissolved in a variety of solvents.

However, highly soluble fungicides can also be more prone to leaching, which means they can be washed away by rain or irrigation and may not be as effective in protecting the plants.

Overall, the formulation of a fungicide is an important consideration in its effectiveness and safety. Careful consideration of the active ingredient, mode of action, solubility, and other ingredients can help to ensure that the fungicide is properly formulated for the specific needs of the plants and the environment in which they are grown.

Fungicide formulation development options are determined by pathogen biology and active ingredient physical-chemical characteristics, including:

Fungal biology – true fungi vs. oomycetes

Active ingredient solubility – hydrophilic or lipophilic

Active ingredient uptake – contact, systemic or translaminar

Application timing – protective, curative or eradicative

Mode of action – spores (protective; respiratory inhibition) or mycelia (curative; ergosterol inhibition)

Fungal Biology

True fungi (Kingdom Mycota) share several biochemical or structural features, which are the targets of fungicides and may impart specificity i.e., the fungicide kills or inhibits fungi.

The fungicidal target Chitin is a component of fungal and insect cell walls, and ergosterol is a component of fungal cell membranes but is absent in animal and plant cell membranes. Ergosterol is formed in fungi via a multistep ergosterol synthesis pathway – a factor that has important implications for resistance management.

The Oomycetes (Kingdom Protista) are not true fungi but use the same mechanisms and strategies to infect plants. Plant pathologists group then with fungal pathogens.

Oomycetes include destructive plant pathogens, including the genus Phytophthora (potato late blight), Pythium (damping off disease) and Peronospora / Pseudoperonospora (Downy Mildews). Instead of chitin, oomycetes cell walls contain cellulose, and ergosterol is not a principal sterol in the Oomycete cell membrane.

Fungicides targeting chitin and ergosterol synthesis are thus generally ineffective against these pathogens. Downy mildews can be confused with unrelated fungal powdery mildews but are insensitive to fungicides able to control powdery mildew.

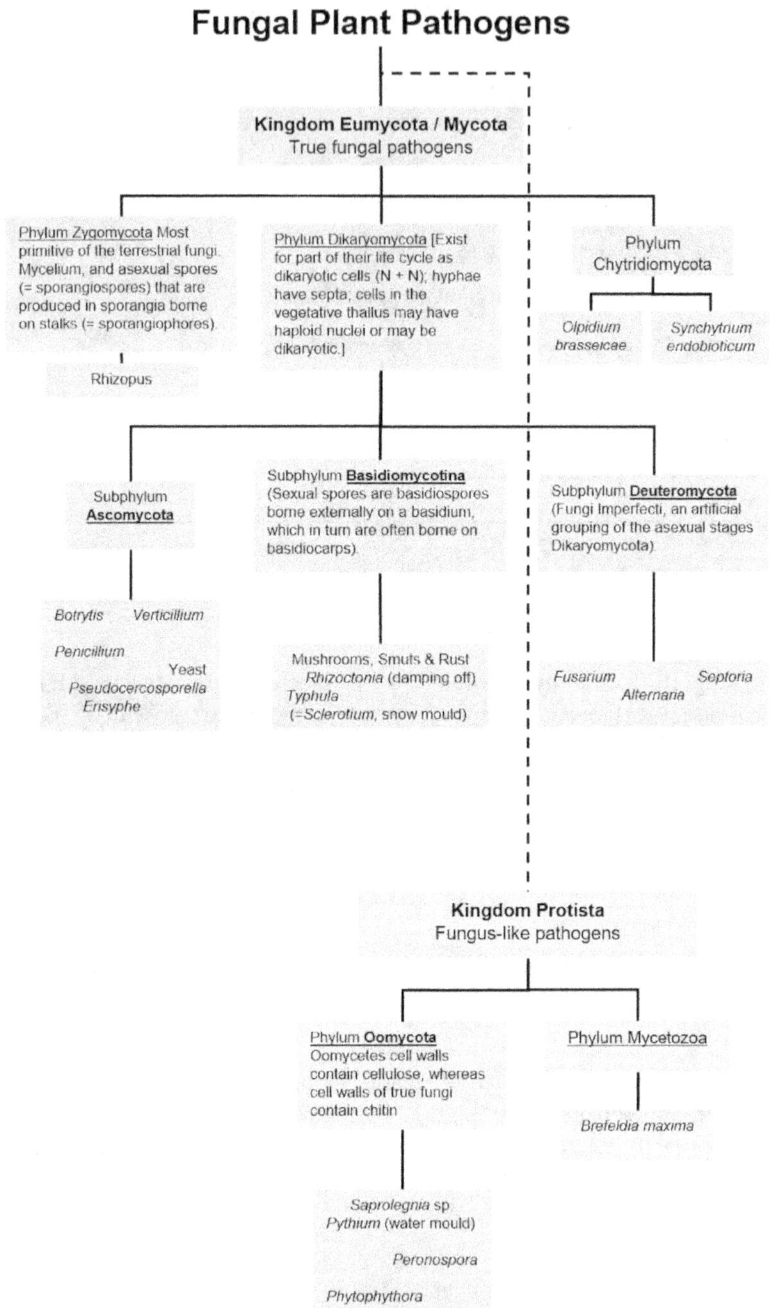

Figure 11.1 Classification of major fungal plant pathogens

When spores land on a leaf, they germinate and form a germ tube before forming an appressorium, which serves as an attachment structure at the penetration site.

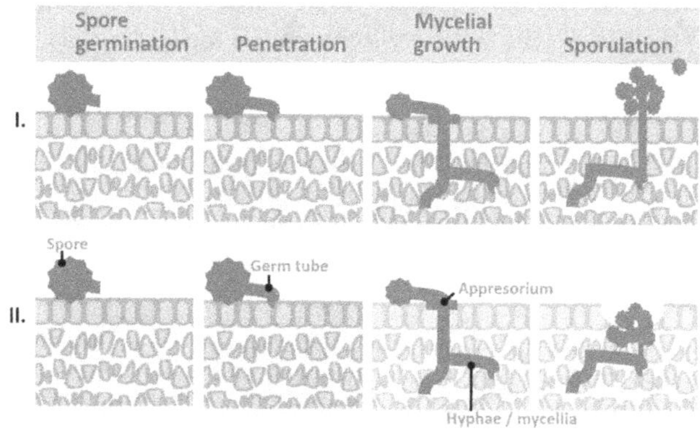

Figure 11.2: Infection and colonization by i) biotrophic and ii) necrotrophic pathogens. Biotrophic pathogens form infection structures to establish compatibility with the host and feed from living plant cells. Necrotrophic pathogens survive by killing plant tissue to obtain nutrients and advance through the colonization of dead tissue.

The penetration hyphae grow through the epidermal cell and intercellular space, first forming primary invading hyphae and then secondary hyphae populating intercellular spaces.

In compatible interactions, primary haustoria form in the mesophyll cells, and within days the formation of secondary haustoria occurs. Two to three weeks after infection, new spores erupt from the leaf – these may be wind or water dispersed and serve as the source of inoculum for new infections.

Fungicide systemicity and activity

With regard to systemicity, fungicides may be classified as contact (non-systemic), translaminar (locally systemic) or systemic.

Preventative or protective (contact) fungicides stop spore germination on the leaf surface to prevent infection. Applied before disease infection, they tend to be more lipophilic as they must remain on the leaf surface to prevent spore germination and penetration.

As contact fungicides tend to be based on older chemistries, they have multisite modes of action (rather than the more specific modes of action of more modern, systemic fungicides. As there is less risk of resistance, there is less need to combine multiple active ingredients, which may be incompatible with each other or with formulation components).

Contact fungicides generally require higher doses than systemic fungicides, and due to their exposure to the environment on the leaf surface, they tend to be less UV- & rainfast, reducing their duration of efficacy to 3-10 days. Contact fungicides thus require more applications than systemic fungicides.

Formulation development options for contact fungicides include increasing rainfastness and UV stability, improving efficacy (reduced rates) through optimized distribution on the leaf surface.

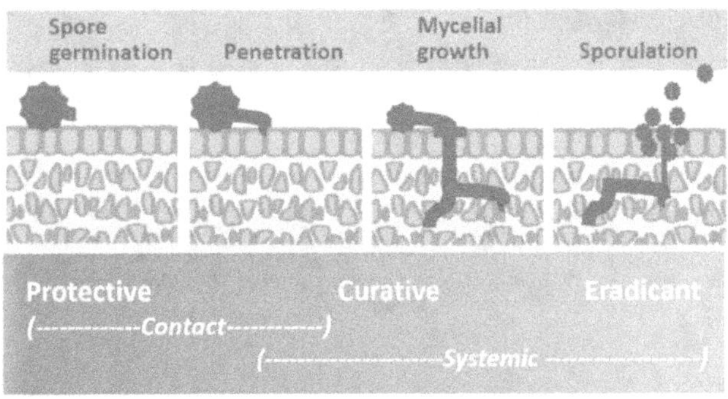

Figure 11.3: Protective, curative and eradicant fungicide timing as a function of disease development.

Curative/eradicant (systemic) fungicides stop early pathogen development inside the plant. Eradicant fungicides are effective at advanced stages of colonization (when symptoms are present).

Applied after infection but before visible symptoms, active ingredients with curative function tend to be less lipophilic as they need to penetrate the leaf and move throughout the plant to target existing infection.

Systemic fungicides are typically effective at lower doses than contact fungicides because they penetrate the plant and are thus more UV- & rainfast, with a duration of efficacy of 14-28d, and require fewer applications. In addition, systemic fungicides may have translaminar activity, allowing them to pass through the leaf from the upper (sprayed) to the lower (unsprayed) surface and to target pathogens on the underside of leaves.

Systemic insecticides are relatively modern, more selective chemistries and tend to have a single-site mode of action. To address resistance issues, they are typically used in combinations with other insecticidal active ingredients, which can introduce compatibility issues that must be addressed by appropriate formulation and adjuvant technologies.

Fungicide systemicity is associated with higher residues (MRLs) and the risk of phytotoxicity through systemic accumulation.

Formulation development options for systemic fungicides include optimizing uptake kinetics to improve efficacy while avoiding phytotoxicity and optimizing formulations for compatibility in tank mixes.

Depending on the formulation choice (EC; SC) and choice of adjuvants, curative activity may be increased (usually at the cost of protective activity) or vice versa.

Translaminar fungicides (locally systemic), such as QoI or Strobilurin fungicides redistribute the fungicide from the upper, sprayed leaf surface to the lower, unsprayed surface. Some Triazole (protective and early curative) fungicides may also have translaminar activity.

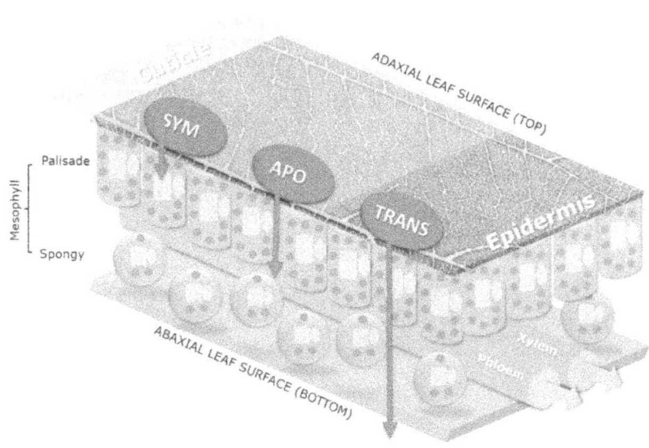

Figure 11.4: APO; apoplastic (extracellular), SYM; symplastic (intracellular) or TRANS; translaminar (from adaxial to abaxial leaf surface) active ingredient pathways.

Translaminar movement takes place via the lipophilic leaf surface and lipophilic extra cellular pathways.

By affecting uptake across the cuticle and intracellular or extracellular transport within the plant, adjuvants are an additional factor deciding whether chemical and biochemical active ingredients have apoplastic (extracellular), symplastic (intracellular) or translaminar (from adaxial to abaxial leaf surface) activity.

. . .

Fungicide Classes and agrobiological groups

Active ingredient physical-chemical characteristics properties may be obtained through various sources, including the Pesticide Properties Database (University of Hertfordshire) and ePesticide Manual (British Crop Protection Council)

It is possible to identify trends and group the key fungicide modes of action based on our understanding of pathogen biology, active ingredient physical-chemical characteristics and fungicidal systemicity.

Mode of action classes may be grouped into systemic or contact fungicides based on whether they are hydrophilic or lipophilic.

Class	Family	AI	logP	Systemicity
Systemic				
Unknown	**Phosphonate**	Fosetyl-al	-2,1	Systemic (xylem, phloem)
DMIs	**Triazoles**	Flutriafol	2,3	Systemic, xylem (protectant, curative, eradicant)
Carboxylic acid amides	**Morpholines**	Fenpropidin	2,9	Systemic, xylem (protectant, curative, eradicant)
SDHI (Succinate dehydrogenase inhibitors)	**Oxathiin carboxamide**	Carboxin	2,3	Systemic (xylem)
	Pyridine carboxamide	Boscalid	2,96	Systemic (xylem) - protective & curative
	Pyrazole carboxamide	Fluxapyroxad	3,13	Systemic - protective & curative
QO, complex III	**Strobilurins**	Azoxystrobin	2,5	Systemic (xylem), protectant, curative, translaminar
DNA synthesis	**Acylalanines; Phenylamide**	Metalaxyl	1,75	Systemic (protectant, curative)
Tubulin inhibition	**Benzimidazoles;**	Thiabendazole	2,39	Systemic with xylem mobility (protectant, curative)
	Thiophanates	Thiophanate	1,5	Systemic with xylem mobility (protectant, curative)
Contact				
DNA and RNA synthesis	**Dicarboximides**	Iprodione	3	Contact, apoplastic (protective, curative)
QI, complex III	**Sulfonamide**	Cyazofamid	3,2	Contact; translaminar (oomycetes)
Glycerol synthesis	**Phenylpyrroles**	Fludioxonil	4,12	Contact (spore germination)
Multi-site activity	**Inorganics**	Copper*	(0,44)*	Contact (protectant)
	Dithiocarbamate	Mancozeb*	(0,26)*	Contact (protectant)
	Phthalimides	Captan	2,8	Contact (protectant)
	Chloronitriles	Chlorothalonil	2,92	Contact (protectant)

Figure 11.5: Fungicide agrobiology groups by class, solubility, and systemicity

Fungicide active ingredients – predictive models for formulation and adjuvant options

By placing fungicide active ingredients into the predictive Bromilow model based on their lipophilicity (logP) and partition coefficient (pKa), we can determine and visualize pesticide mobility and predict their suitability for contact and systemic activity.

Many common fungicides are clustered around the "low" end of the systemicity scale and are primarily xylem mobile.

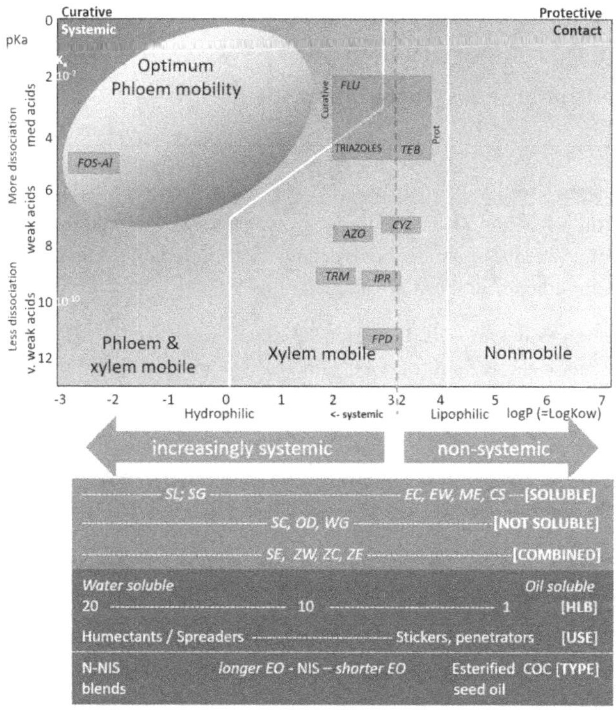

Figure 11.6: Expanded Bromilow diagram for predicting Fungicide application and systemicity, as well as formulation types and adjuvants for hydrophilic and lipophilic active ingredients (Fungicides: FOS-Al fosetyl-aluminum FLU flutriafol TEB tebuconazole FLP fluazifop CYZ cyazofamid AZO azoxystrobin TRM thiram IPR iprodione FPD fenpropidin)

For a fungicide to be considered curative, the active ingredient must be systemically translocated within the plant to inhibit (cure) existing fungal hyphal growth within the leaf. Very few fungicides are phloem systemic (FOS-Al; CAR) and truly curative, and these may translocate to leaf tips and meristems, reducing their efficacy and increasing the risk of phytotoxicity.

Protective fungicides are described as having contact action, remaining on the leaf surface where they can protect the plant against the germination of newly landed fungal spores.

The expanded Bromilow model can be used to predict appropriate formulations and adjuvants for specific active ingredients.

For hydrophilic active ingredients, solubilization in the aqueous spray solution in water-based formulations is generally possible, while for lipophilic active ingredients that are soluble in organic solvents such as oil, emulsion formulations or microcapsule formulations are appropriate. If multiple active ingredients are to be mixed, SE, ZW, ZC or ZE can be considered.

Formulation options for systemic fungicides

Surfactants - wetters Surfactants allowing spray droplets to spread beyond their initial contact area, providing a greater area of uptake for systemic active ingredients, or a more homogenous distribution on the leaf surface for active ingredients with contact activity.

Surfactants with a low HLB (shorter EO chains) are absorbed into the cuticle and increase the fluidity of waxes, increasing the permeance of lipophilic fungicides in the cuticle (possible crop phytotoxicity).

Crop oils and COCs Crop oils and crop oil concentrates (COCs) enhance spreading and penetration of some (high MW, more lipophilic) systemic and translaminar fungicides. Can induce phytotoxicity.

Humectants Enhance the hydration of the cuticle and the permeance of hydrophilic fungicides into the cuticle.

Formulation Emulsions tend to be more effective than suspensions; suspension formulations often allow higher loading and are less phytotoxic.

Hydrophilic surfactants include oil-in-water emulsifying agents, detergents, and solubilizing agents and are suited for foliar-applied hydrophilic fungicides.

Lipophilic surfactants include penetrators (oils) and are suited for foliar-applied lipophilic active ingredients.

Formulation options for contact fungicides

Surfactants - wetters Surfactants allowing spray droplets to spread beyond their initial contact area, providing a more homogenous distribution on the leaf surface for active ingredients with contact activity.

Stickers increase droplet or particle adherence to the leaf surface and prevent run-off on application and wash-off during heavy rainfall. Spreader-stickers contain surfactants (spreading) and stickers (increase adhesion)

Crop oils and COCs enhance spreading of contact fungicide s and can improve rainfastness but can lead to leaf penetration and phytotoxicity.

Formulation Suspension concentrate formulations are generally retained better than wettable powder formulations.

Optimization for tank-mixed fungicides

The need to incorporate agricultural actives (ais) in high electrolyte systems, such as fertilizers for in-furrow applications or in tank mixes with other fungicides, insecticides, and biostimulants for spray applications, provides a significant challenge to the formulator, including flocculation (precipitation) and the risk of blocked nozzles, phytotoxicity, active ingredient degradation and the inactivation of biologicals.

Precipitation may occur in high-electrolyte solutions, especially SC formulations. Most non-ionic adjuvants have low cloud points and phase-separate under high-salt conditions; thus, cationic, anionic, or, increasingly, amphoteric surfactants are required. However, these adjuvants are often incompatible with other actives, may be environmentally or plant-toxic, or cause excessive foaming.

Phytotoxicity may occur when mixing active ingredients, such as cytotoxic sulfur, copper, and captan fungicides mixed with oil-containing products (as they are cytotoxic; leaf penetration can lead to phytotoxicity).

Phosphites (phosphorous acids) and copper combinations can be phytotoxic, as copper can become more available and phytotoxic in acidic solutions.

Triazole fungicides mixed with non-ionic surfactants or foliar fertilizers can lead to increased uptake, with increased systemic accumulation and phytotoxicity.

Antioxidants, herbicide safeners, and the judicious choice of adjuvants can help alleviate the risk of phytotoxicity, while evidence indicates that the co-application of certain insecticides may improve crop compatibility.

Degradation occurs when fungicidal active ingredients degrade each other in tank mixes, for example, lime in mixes with Captan, Mancozeb, Iprodione (all susceptible to alkaline hydrolysis).

Formulation optimization options include the inclusion of alkalizers or acidifiers, and compatibilizer adjuvants (amphoteric surfactants and hydrotropes).

Adjuvants, such as combinations of amphoteric surfactants and hydrotropes, that work as compatibilizers allow different active ingredients to be mixed to increase the effectiveness of crop protection formulations.

When pH is low, amphoteric surfactants have cationic characteristics and when pH is alkaline have anionic characteristics.

Hydrotropes are alternatives to common surfactants and improve the solubility of hydrophobic particles without forming micelles.

Inactivation of biologicals, such as copper (cytotoxic, antibacterial) with bacterial biofungicides and fungicides, such as triazoles with fungal biofungicides (e.g., Trichoderma) is an issue when tank-mixing, and this inactivation can be alleviated using compatibilizer adjuvants, as well as formulation compartmentalization (SE, OD, CS, inclusion technologies).

Overall, the formulation of a fungicide is an important consideration in its effectiveness and safety. Careful consideration of the active ingredient, mode of action, solubility, and other ingredients can help to ensure that the fungicide is properly formulated for the specific needs of the plants and the environment in which they are grown.

V

BIOPESTICIDE MODE OF ACTION AND FORMULATION

12

INTRODUCTION

Biopesticides are pest control products that are derived from natural materials and are used to control or prevent pests in crops, plants, and animals. They are an important alternative to traditional chemical pesticides, as they are often safer for human health and the environment.

Biopesticides can be classified into several types based on their mode of action, including biological control agents, pheromones, and plant-derived compounds. In this section, we will discuss the various types of biopesticides and their modes of action. We will also explore the advantages and disadvantages of biopesticides, and we will consider the factors that influence their performance and effectiveness.

Formulating biopesticides into effective and stable products is a complex process that involves the selection of appropriate active ingredients, adjuvants, and excipients.

In this chapter, we will discuss the various types of biopesticides and their modes of action, and we will explore the factors that are considered when formulating biopesticides. We will also discuss the challenges and considerations involved in formulating biopesticides for different applications and environments.

Biopesticides are regulated by government agencies to ensure that they are safe for human health and the environment. These regulations vary by country and can be complex, as biopesticides are often derived from natural materials and can have different modes of action than traditional chemical pesticides. In this chapter, we will discuss the regulatory frameworks that govern the development, registration, and use of biopesticides.

We will explore the various agencies and guidelines that are involved in the regulatory process, and we will discuss the criteria that are used to assess the safety and efficacy of biopesticides. We will also discuss the challenges and considerations that are involved in complying with biopesticide regulations, and we will consider the impact of these regulations on the development and use of biopesticides.

13

A BIOPESTICIDE TIMELINE

The application of biopesticides in plant protection is certainly not a new innovation – a few examples are presented in the following:

The efficacy of biopesticides is - to a large extent - driven by inducible plant defense responses evolved over more than 100 million years. These defense responses have proven to be sufficiently effective to permit the evolution of a complex flora.

Crop pests and diseases, as well as the activity of beneficial insects have been known since the beginning of agriculture, some 10K years ago.

In 1835, Agostino Maria Bassi identified the ascomycetes fungus *Beauveria bassiana* as a disease of silkworm. By the turn of the nineteenth century, scientists were reporting on the biology and pathology of entomopathogenic fungi.

In 1901 a bacterial pathogen of silkworms was discovered, subsequently named *Bacillus thuringienses* (Bt).

Field trials with Bt to control the European corn borer were conducted in the late 1920s, and in 1938 the first commercial Bt preparation was introduced.

Today, entomopathogenic bacteria are widely used to control insect pests, typically via their digestive tracts, while Bt-toxin crystals are the basis of microbial-derived bioinsecticides which are activated upon ingestion by lepidopteran insect pests, leading to cell disintegration and insect death.

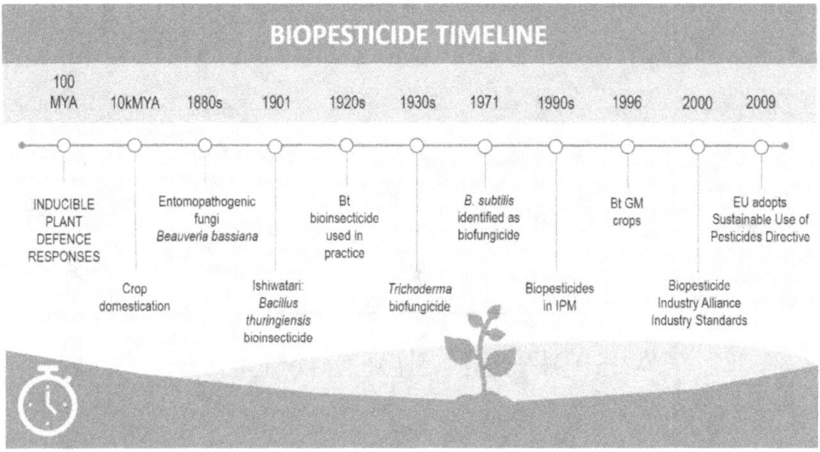

Figure 13.1: Biopesticide Timeline.

In the 1930s, several species of the soil-borne Ascomycetes fungus Trichoderma began to be used as biocontrol agents against fungal diseases of plants, through the mechanisms of antibiosis, parasitism, plant-defense induction, and competition.

In 1971, the endospore-forming bacterium *Bacillus subtilis* was identified as a biofungicide able to suppress seed and seedling disease.

Bacillus strains such as *B. subtilis, B. amyloliquefaciens and B. licheniformis* produce antimicrobial metabolites such as lipopeptides, which exhibit a broad antimicrobial spectrum and surfactant activity.

Today, B. subtilis and its variants are among the most-used biofungicides to control soil and seed-borne pathogens, as well as foliar plant diseases.

With the increased focus on Integrated Pest Management (IPM) in the 1990s, interest in genetic modification (GM) and biopesticide development grew, and in 1996 the first insect-resistant Bt GM crops were introduced.

Bacillus thuringiensis functions as an insecticide when its toxin (Bt toxin) is directly incorporated into plants through the use of genetic engineering.

The increased public awareness of the benefits of Integrated Pest Management strategies, which combine the use of biopesticides with natural predators, pest-resistant varieties, cultivation techniques and conventional crop protection chemistries is a key factor driving the growth of the biopesticide market.

Reflecting the increased contribution of biological strategies in crop protection, the Biopesticide Industry Alliance (changed to Biological Products Industry

Alliance in 2017) was initiated in 2000 to ensure Industry Standards and increase awareness of biopesticides and biologicals in agriculture.

In 2009, the EU adopted the Sustainable Use of Pesticides Directive (2009/128/EC) promoting the implementation of biopesticide-based integrated pest management solutions.

14

BIOHERBICIDE MODE OF ACTION

Bioherbicides include microorganisms that infect weeds, and natural compounds with herbicidal activity sourced from microorganisms or plants.

Bioherbicides fall into two major classes:

- Microbial bioherbicides
- Bio-derived (biochemical) bioherbicides

Bioherbicides target many of the same plant metabolic processes targeted by conventional herbicides, but they may bind differently to enzymes, or they may have multiple modes of action, making them potentially useful for treating herbicide-resistant weeds.

Microbial Bioherbicides

Microbial bioherbicides include *bacterial and fungal endophytes* (microorganisms which colonize plants without causing disease) as well as viruses.

The modes of action of fungal bioherbicides (*mycoherbicides*) include the action of plant cell wall degrading enzymes, photobleaching molecules (macrocidins) and phytotoxic molecules such as the diterpene chenopodolin, and growth-regulating plant hormones such as indole acetic acid, IAA. *Phoma macrostoma* is an example of a mycoherbicide developed for pre-emergence weed control.

Bacterial bioherbicides exhibit phytotoxic activity attributed to specific metabolites as well as the elicitation of plant defense responses, such as programmed cell death through the induction of reactive oxygen species.

Bio-Derived (Biochemical) Bioherbicides

Bio-derived (biochemical) bioherbicides (phytotoxins) are based on mechanisms used by microbes and plants to exclude invasive plant species from their immediate vicinity.

Allelopathy is a mechanism by which plants produce one or more biochemicals (*allelochemicals*) that affect the growth and survival of other plants.

The study of allelopathy has, among others, led to the identification of leptospermone, a herbicidal allelochemical produced by the Lemon Bottlebrush, and the subsequent development of its commercial chemical analogue, Mesotrione (Callisto).

Biochemical bioherbicides target many of the same plant metabolic processes targeted by conventional (synthetic) herbicides. However, natural phytotoxins may bind to proteins at different sites than their synthetic counterparts, or may have multiple modes of action.

For plants that produce phytotoxins, an autotoxicity avoidance mechanism is needed to prevent translocation of potentially toxic root exudates to sensitive parts of the plant.

For example, sorgoleone (from *Sorghum bicolor*), a D1-protein-binding inhibitor of photosystem II can control weeds resistant to the QB-site PSII-inhibitor, atrazine.

Besides its inhibition of PSII electron transport and the concomitant increase in reactive oxygen species (ROS), sorgoleone inhibits the enzymatic activity of thylakoidal HPPD (*hydroxyphenylpyruvate dioxygenase*) and thus the formation of vitamin E, depriving the plant of one of its primary antioxidant protective mechanisms. Finally, sorgoleone inhibits *mitochondrial respiration* by blocking proton-gradient-driven ATP synthesis.

Chloroplast-based metabolic processes inhibited by natural phytotoxins include photosynthetic electron transport, chlorophyll and carotenoid biosynthesis, amino acid biosynthesis as well as lipid membrane function and stability.

Sorgoleone (*Sorghum bicolor*) and stigmatellin (from *Stigmatella aurantica*) are inhibitors of photosystem II (and *hydroxyphenylpyruvate dioxygenase*, HPPD), while pyridazocidin (*Streptomyces* spp.) is an inhibitor of photosystem I.

Aurachins (*Stigmatella aurantica*) inhibit electron flow between PSII and PSI.

In photosynthesis, the photolysis of water leads to an accumulation of protons in the thylakoids stroma which drive the synthesis of ATP (photophosphorylation) via the enzyme *ATP synthase* (ATPase).

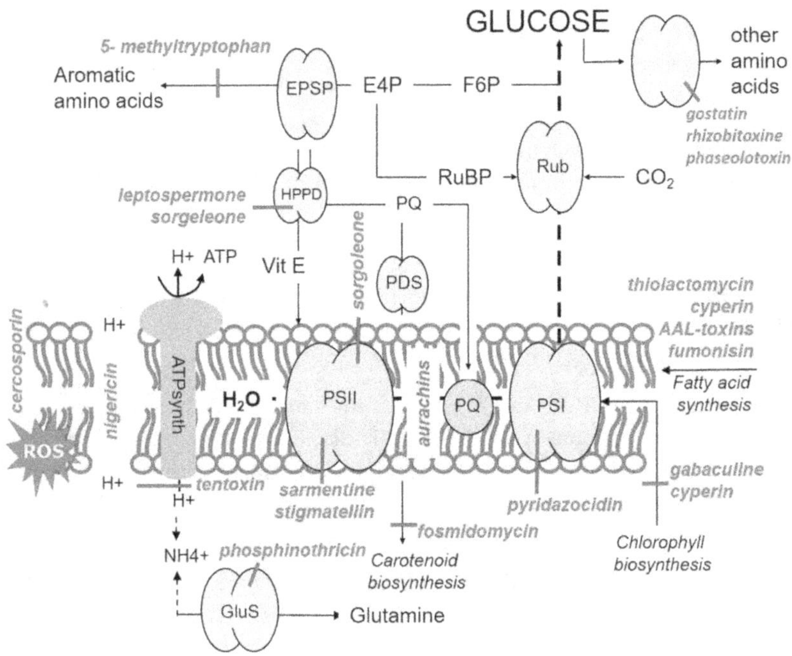

Figure 14.1: Target sites of natural phytotoxins affecting metabolic processes within the thylakoid membrane.

Nigericin (*Streptomyces*) uncouples the proton gradient, inhibiting photophosphorylation while tentoxin (*Alternaria tenuis*) blocks photon flow through the chloroplast ATPase.

Gabaculine (*Streptomyces toyacaenis*) and cyperin (*Preussia fleischhakii*) are natural phytotoxins inhibiting chlorophyll biosynthesis. Carotenoid biosynthesis may be blocked at different sites by the action of fosmidomycin (*Streptomyces lavendulae*).

Leptospermone (*Leptospermum scoparium*) and sorgoleone inhibit the hydroxyphenylpyruvate dioxygenase (HPPD) enzyme, blocking both carotenoid biosynthesis and the formation of the antioxidant protectant, vitamin E.

Like vitamin E, carotenoids have a protective antioxidant function through their quenching or scavenging of oxygen radicals.

While Leptospermone does have herbicidal effects, the rate needed for commercial levels of efficacy is too high to be practical. Leptospermone analogues were too toxic, persistent or non-selective, and consequently synthetic triketone HPPD-inhibitor herbicides - such as Mesotrione (Callisto) - have been successfully developed.

Phytotoxins such as sarmentine (*Piper longum*) and cercosporin (*Cercospora kikuchii*) lead to loss of membrane integrity through *reactive oxygen species* (ROS), arising because of the over-reduction of the photosynthetic apparatus due to PSII inhibition.

The repair and maintenance of lipid membranes may be inhibited by herbicidal phytotoxins such as thiolactomycin (*Streptomyces* spp.) and cyperin (*Preussia fleischhakii*).

Biosynthesis of the *aromatic amino acid* tryptophan may be inhibited by the phytotoxin 5- methyltryptophan (*Cantharellus cibarius*), while a range of phytotoxins such as gostatin (*Streptomyces*), rhizobitoxine (*Bradyrhizobium*) and phaseolotoxin (*Pseudomonas*) inhibit the biosynthesis of other amino acids.

Phosphinothricin (*Pseudomonas syringae*) is a herbicidal phytotoxin inhibiting the synthesis of the amino acid glutamine.

Mitochondria-based metabolic processes inhibited by natural phytotoxins include respiration and lipid biosynthesis.

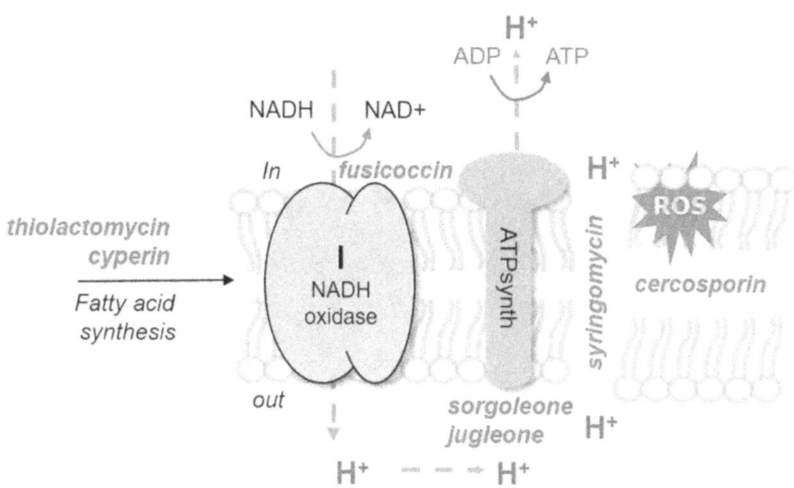

Figure 14.2: Target sites of natural phytotoxins affecting metabolic processes within the mitochondria.

Respiration can be blocked through the inhibition of NADH oxidation by fusicoccin (*Fusicoccum amygdali*). Syringomycin (*Pseudomonas syringae*) uncouples the proton gradient through the formation of membrane pores, while sorgoleone (*Sorghum bicolor*) and juglone (*Juglans* spp.) block proton-driven ATP synthesis via the H^+ ATPase. *Fatty acid biosynthesis* in the mitochondria can be inhibited by thiolactomycin (*Streptomyces*) and cyperin (*Preussia fleischhakii*).

Mitochondrial respiratory electron transport processes are evolutionarily related to photosynthetic electron transport, and herbicides affecting photosynthetic electron transport may affect respiration in mammals, insects and microorganisms. This could have implications for the environmental impact of this group of pesticides, and is the subject of growing research interest.

Thaxtomin (*Streptomyces scabies*) is a microbial phytotoxin that controls grass, broadleaf and sedge weeds by disrupting *cellulose biosynthesis.*

Additional modes of action for bioherbicides include *gene expression and regulation,* as well as effects on hormonal processes such as *jasmonate mimicking* and *auxin signalling*.

15

BIOINSECTICIDE MODE OF ACTION

Bioinsecticides include microorganisms that infect insects, and natural compounds with insecticidal activity sourced from microorganisms, plants or animals.

Bioinsecticides fall into three major classes:

- *Microbial bioinsecticides*
- *Plant-incorporated protectants* - a form of biological insecticide in which foreign DNA such as the BT-toxin gene is inserted into crop genetic material.
- *Bio-derived (biochemical) bioinsecticides* - naturally occurring compounds, typically produced by plants for defense from predation, but also include microbial extracts that control insects or induce plant defense responses.

Bioinsecticides known not to harm non-target (beneficial) insect species may be considered for use during flowering, a period where many conventional pesticides cannot be used.

Microbial insecticides

Microbial insecticides include *entomopathogenic fungi,* capable of acting as pathogenic parasites of insects (and other invertebrates such as nematodes), and typically belong to the divisions Ascomycetes and Zygomycetes.

Besides their parasitic action, entomopathogenic fungi may also produce fungal toxins.

Beauveria bassiana is one of the most studied entomopathogenic fungal species, and its mode of action involves several bioactive metabolites, including cuticle-degrading enzymes and toxic bioactive molecules.

Bassiacridin is a metabolite which affects the epithelial cells of tracheae, air bags, and integuments in insects.

Beauvericin is a metabolite with antibiotic and insecticidal effects. Although primarily associated with Beauveria, it is also produced by several other fungi, including several Fusarium species and it may thus occur in cereal crops contaminated with these fungi.

Beauvericin is active against Gram-positive bacteria (a factor to note if mixed with Bacillus-based biopesticides) and mycobacteria, and may also induce programmed cell death in mammals.

Entomopathogenic bacteria such as *Bacillus thuringiensis* (BT) are used to control insect pests, typically via their digestive tracts. Some strains of *Bacillus thuringiensis* (Bt) are specific to Lepidopteran species, while other strains may target larvae of Diptera and Hemiptera.

Reports of Bt toxicity toward non-target organisms, as well as binding of the bacterial toxin to clay soils have been the topic of debate, most researchers agree that Bt monitoring must continue, and their use modified as appropriate.

Entomopathogenic viruses are usually specific to individual species: *Cydia pomonella* granulovirus specifically targets the codling moth *C. pomonella* (a major pest of fruit trees) killing the insect at the larval stage.

Beneficial nematodes (*e.g. Steinernema*), while not microorganisms, are sometimes classified as such and are used in the biological control of insect pests.

Plant-incorporated protectants

Plant-incorporated protectants (PIPs) are a form of biological insecticide in which foreign DNA is inserted into crop genetic material (GM crops).

Bacillus thuringiensis toxin genes may be directly incorporated into plants using genetic engineering, giving these plants tolerance primarily to Lepidopteran insect pests, such as caterpillars and beetles.

This specificity is due to the fact that different strains of Bt are specific to different receptors in the insect gut wall, and damage to the gut occurs following ingestion and binding to a gut wall receptor.

RNA interference (RNAi) is an emerging technology, whereby RNA molecules capable of disrupting genes conferring tolerance to insecticides may be used to target resistant insects such as Potato Beetles.

Bio-derived (biochemical) insecticide

Bio-derived (biochemical) insecticides are naturally occurring compounds, typically produced by plants for defense from predation, but also include microbial extracts that control insects or induce plant defense responses. Insect *pheromones* are used to attract insect pests to traps, or to disrupt mating cycles.

BT toxin crystals are microbial-derived insecticidal proteins which are activated upon ingestion by lepidopteran insect pests, leading to cell disintegration and insect death.

Surfactin, a biodegradable lipopeptide biosurfactant isolated from Bacillus species (see below) has mosquitocide activity, probably through its surfactant properties.

Chitosan is a polysaccharide derived from fungi or crustacean shells, capable of inducing systemic resistance against insect pests.

Insect pheromones are used to attract insect pests to traps, or to disrupt mating cycles.

Azadirachtin (a botanical insecticide extracted from natural neem oil) acts as an insect growth regulator, anti-feedant, repellent, sterilant, and oviposition inhibitor

Plant extracts such as *garlic extracts or canola oil* can repel insects, or act as antifeedants, while other plant extracts such as *Nicotine* (isolated from tobacco), *Pyrethrins* (isolated from Chrysanthemum plants) or *Ryanoids* (isolated from *e.g. Ryania speciosa*) have neural insecticidal activity, leading to hyperstimulation -or paralysis - and death. For a more detailed summary of neural insecticidal activity, please refer to Chapter 2.

In *Cholinergic* (ACh) neurons, *Nicotine* competes with ACh neurotransmitters and competitively binds to the ACh receptor on the postsynaptic nerve, resulting in uncontrolled nerve stimulation (hyperstimulation) and death.

Pyrethrins disrupt normal nerve function by binding to voltage-gated sodium channels, preventing them from closing normally and giving rise to neural hyperactivity and insect death.

Spinosad is a bioinsecticide based on natural metabolites, and is produced by fermentation of naturally occurring bacteria (*Saccharpolyspora spinosa*).

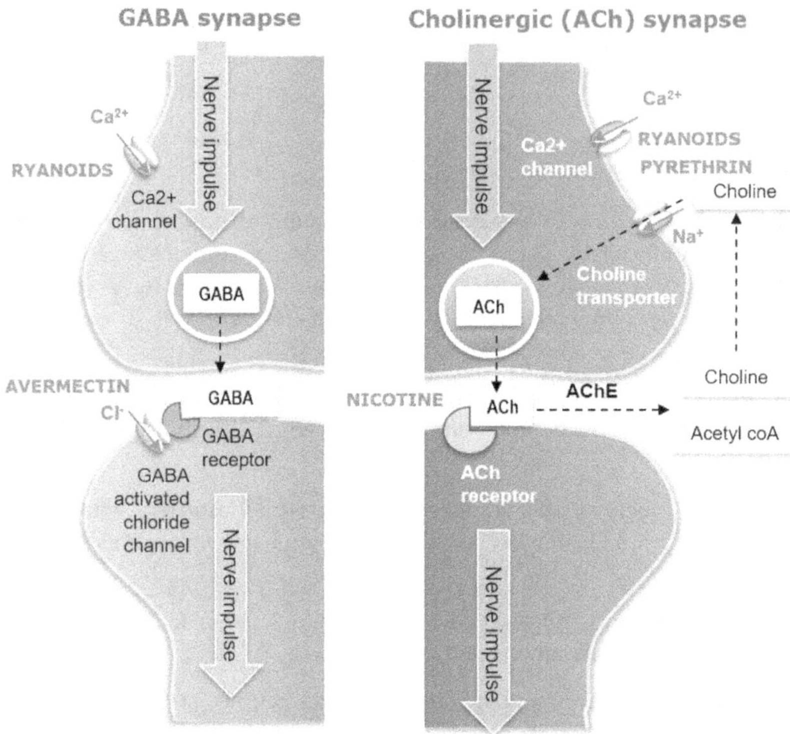

Figure 15.1: Mode-of-action of microbial and plant-derived neural bioinsecticides.

Spinosad is highly active, by both contact and ingestion, in numerous insect species, and its mode of action targets the insect nervous system.

Its mode of action targets binding sites on the nicotinic acetylcholine receptors that are distinct from those at which other insecticides have their activity, and Spinosad binding disrupts acetylcholine neurotransmission.

Spinosad also has secondary effects as a γ-amino-butyric acid (GABA) neurotransmitter agonist [*agonists* activate the receptors to which they bind, increasing the inhibitory or excitatory effect of neurotransmitters] and kills insects by hyperexcitation of the insect nervous system.

Avermectins (naturally occurring compounds generated as fermentation products by the soil bacteria, *Streptomyces avermitilis*) affect insect-specific glutamate-gated chloride channels in GABA neurons, causing an influx of chloride (Cl^-) ions into the postsynaptic neuron - leading to hyperpolarisation, paralysis and death.

Some insecticides have a high specificity towards insects, due to their greater affinity for insect neuroreceptors, relative to neuroreceptors in mammals.

In contrast to specific insecticides, other insecticides may target metabolic processes which are common to insects and mammals.

As the AChE enzyme is similar in insects and mammals, this group of insecticides is not considered selective, and appropriate care must be exercised in their application.

Insecticides affecting glutamate-gated chloride channels are specific to insects, as these channels do not exist in mammals.

Ryanoids are insecticidal alkaloids which interact with ryanodine calcium-channel receptors in insects, resulting in paralysis and death.

Other plant extracts such as *Rotenone* can inhibit mitochondrial electron transport (respiration).

Finally, bio-derived insecticides include naturally occurring desiccants, such as borate or silica (*diatomaceous earth*, the fossilized remains of aquatic diatoms) which absorb the oil and fats from the cuticle of the insect's exoskeleton, leading to desiccation and death.

It should be noted that not all natural products are able to be registered as biopesticides.

For example, despite their natural origins, pyrethrins, spinosad and abamectin are registered as conventional (chemical) pesticides due to their neurotoxic mode of action.

Azadirachtin is a botanical insecticide and Spinosad and Abamectin are naturally occurring fermentation products. Both Azadirachtin and Spinosad are certified for organic production (at least in the US), but not Abamectin.

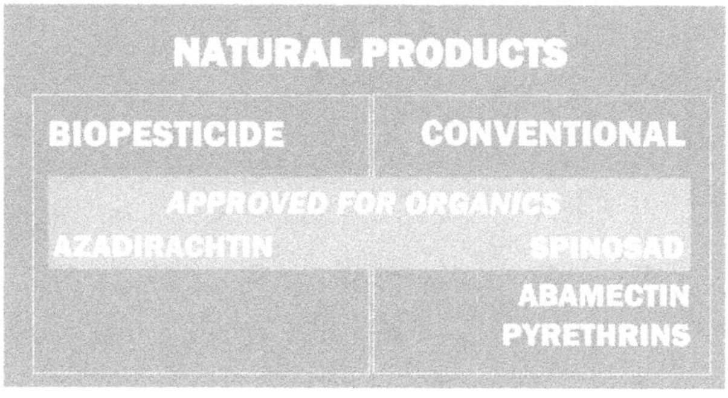

Figure 15.2: Natural products may be registered as biopesticides or conventional pesticides.

This discrepancy appears to be related to human and aquatic toxicity issues for Abamectin. Spinosad and Abamectin are not considered biopesticides, due to their neurotoxic mode of action (whereas Azadirachtin - acting as an insect growth regulator, anti-feedant, repellent, sterilant, and oviposition inhibitor) - is.

16

BIOFUNGICIDE MODE OF ACTION

Biofungicides are formulations of living organisms or natural metabolites used to control the activity of plant-pathogenic fungi.

Biofungicide modes of action include:

---

antibiosis - inhibition of spore germination or pathogen growth

plant defense induction

exclusion - competition and displacement of plant pathogenic fungi by microbial biofungicides

mycoparasitism - parasitization of pathogenic fungi by microbial biofungicides

biostimulation - stimulation of plant growth and resilience

Many successful microbial biofungicides can colonize and grow in treated soil or on plant leaves, providing a long-term source of microbial and biochemical modes of action.

Antibiosis

The genus Bacillus comprises different strains which are commercially available as biofungicides.

Bacillus strains such as *B. subtilis, B. amyloliquefaciens* and *B. licheniformis* produce antimicrobial metabolites such as lipopeptides, which exhibit a broad antimicrobial spectrum and surfactant activity.

Due to their lipid-protein structure, *lipopeptides* can penetrate lipid membranes such as the cytoplasmic membrane of cells, affecting their integrity and ion permeability.

The *Iturin* and *Fengycin* families of lipopeptides exhibit strong antifungal activity across a relatively broad range of pathogenic fungi, with fengycins having activity against Oomycetes such as *Pythium* and *Phytophthora.*

Another family of lipopeptides, *Surfactins* exhibit antiviral and antibacterial characteristics. Despite having low inherent antifungal activity, surfactins may act synergistically in combination with iturin.

Surfactins also function as biosurfactants, acting both as a wetter (by decreasing the surface tension of water droplets) and detergent, and are involved in biofilm formation and motility, while inhibiting the biofilm formation of phytopathogenic bacteria, interfering with their attachment to host plants.

Streptomyces is another naturally occurring gram-positive bacteria. It is found in soil, and can produce antimicrobial compounds (the antibiotic *Streptomycin*).

Localized or systemic plant defense responses

Plants detect invading pathogens and induce defense responses, such as the production of toxic chemicals and cell destruction (hypersensitive responses), to hinder disease development.

Biotic and abiotic plant response mechanisms include chemical-, temperature- and photoreceptors, proximity sensors and accelerometers, and may be compared to smartphone sensors.

Plants mediate defense strategies through complex communications systems, which may be compared to 3G (volatile plant hormones) and Wi-Fi (root-root and root-mycorrhiza interactions) .

Plant defense responses are not initiated until pathogens are detected, due to the energy costs of their production and maintenance. Modern biopesticide strategies must thus consider the kinetics of defense response induction *prior* to pathogen attack.

These strategies include priming and activating plant defense responses prior to pathogen attack, through the application of microbial biofungicides, or through microbial metabolites such as *Harpin,* a natural protein found in common pathogenic microorganisms.

Figure 16.1: Plant response receptors compared to smartphone sensors.

Systemic Acquired Resistance (SAR) describes plant response to pathogen attack through priming of plant defense responses and compounds.

Systemic Acquired Resistance is translocated within plants via the *salicylic acid* (SA) dependent signalling pathway.

Elicitors of plant defense responses registered in the EU and capable of inducing systemic resistance against insect pests include *chitosan* (a polysaccharide derived from fungi or crustacean shells), *laminarin* (a polysaccharide extracted from the seaweed *Laminaria digitate*) and *Saccharomyces* extracts.

Plants utilize pattern recognition receptors (PRRs) to identify specific microbe-associated molecular patterns (MAMPs) from e.g. chitin and its hydrolyzed metabolite *chitosan,* to elicit plant defense responses, including the production of *phytoalexins* (antimicrobial or antioxidative plant metabolites that accumulate at areas of pathogen infection) when applied before or during the inoculation of plants with pathogenic fungi.

Jasmonic acid (JA) mediates the systemic induction of disease defense compounds, including *pathogenesis related proteins* (PR-Proteins) such as *chitinases* and *peroxidases,* which degrade fungal cell walls, and antifungal phytochemicals such as *phenolics, terpenoids* and *alkaloids.*

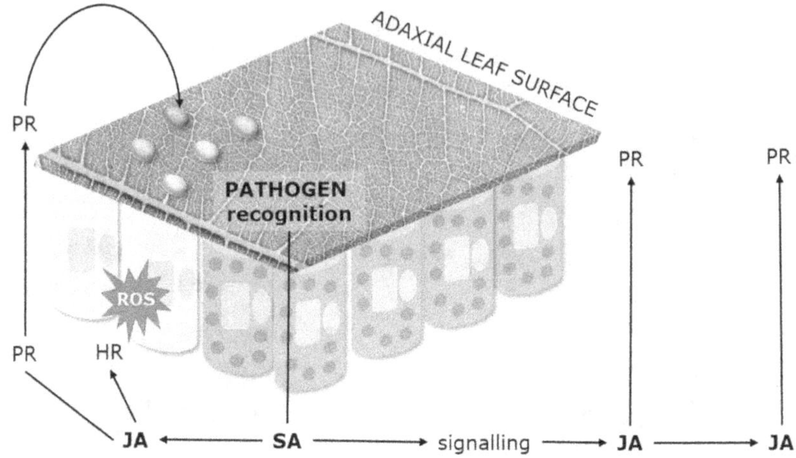

Figure 16.2: Mechanisms of Systemic Acquired Resistance (SAR) as plant response to pathogen attack.

Besides direct antifungal strategies, plants may activate a *hypersensitive response* (HR), also called "programmed cell death," which limits the access of biotrophs (pathogens that cannot survive in a dead host and feed on living tissue) to water and nutrients.

The *hypersensitive response* is a mechanism used by plants to restrict pathogen development by inducing the localized, rapid death of cells in the region immediately surrounding an infection.

Following recognition of the pathogen, the activation of resistance genes triggers an oxidative burst of *reactive oxygen species* (ROS).

Reactive oxygen species draw electrons from chloroplast thylakoid membranes (lipid peroxidation), leading to the loss of chlorophyll and associated chlorotic lesions.

Biostimulation

Biostimulation and protection against biotic (Weeds, Pests and Diseases) and abiotic stresses (including Cold and Heat stress, as well as Drought, Salinity and UV stress) have been reported for many agricultural biologicals, including biopesticides, by improving plant growth and root development (improving drought resistance and nutrient uptake).

Increased crop vigor ensures optimal natural crop defense responses.

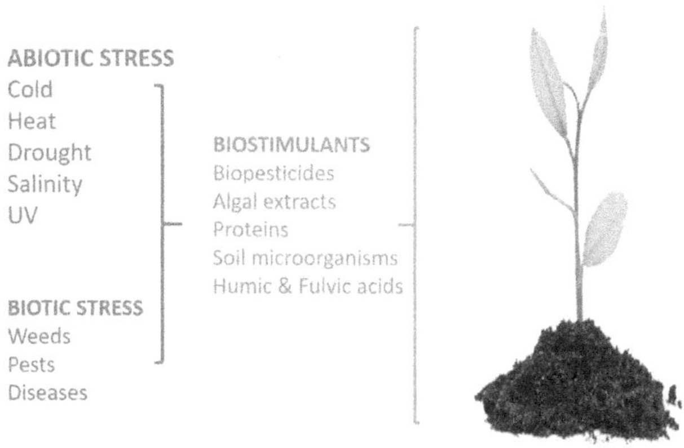

Figure 16.3: Improved crop vigor may be induced by biotic- and abiotic stress, and biostimulants.

Exclusion and parasitism

Microbial biofungicides may compete with plant pathogens for nutrients and infection sites on leaf surfaces and in the soil. Some microbial biofungicides metabolize root exudates that stimulate pathogen germination, reducing their ability to infest plants.

Microbial biofungicides such as *Trichoderma* (a genus of soil fungi) may attack and suppress plant pathogens through a combination of displacement and the production of antimicrobial compounds such as cell-wall degrading enzymes.

17

BIOPESTICIDE FORMULATION

Similar to conventional pesticides, formulation of biopesticides comprises the blending of a biologically active metabolite or microorganism with a carrier or solvent as well as adjuvants, to develop a product which can be delivered to the target weed, pathogen or pest using existing equipment.

The objectives of biopesticide formulation are to stabilize the active metabolite or microorganism, to optimize delivery to the target, to protect the biopesticide during storage and in the field, and to enhance the efficacy of the biopesticide.

Microbial biopesticides

Microbial biopesticides are biological control agents that are used to control pests in agriculture, forestry, and public health. These agents are typically derived from microorganisms, such as bacteria, fungi, and viruses, and can be used as an alternative to chemical pesticides.

The formulation of microbial biopesticides involves several steps, including the selection and isolation of the microbial strain, the production of the microbial culture, and the formulation of the final product.

Selection and isolation of the microbial strain:

The first step in the formulation of microbial biopesticides is the selection and isolation of the microbial strain. This involves identifying microorganisms that have a natural ability to control pests and selecting the most effective strain for use as a biopesticide. The selected strain should be able to survive in the target environment and should have a high level of pest control efficacy.

Production of the microbial culture:

Once the microbial strain has been selected and isolated, the next step is to produce a large quantity of the culture. This is typically done through fermentation, where the microorganism is grown in a controlled environment using a nutrient-rich medium. The microbial culture is then harvested and concentrated to create the final product.

Formulation of the final product:

The final step in the formulation of microbial biopesticides is the formulation of the product. This involves mixing the concentrated microbial culture with a suitable carrier, such as water or oil, to create the final product.

The product may also be formulated with additional ingredients, such as surfactants or stabilizers, to improve its stability and effectiveness.

In conclusion, the formulation of microbial biopesticides involves several steps, including the selection and isolation of the microbial strain, the production of the microbial culture, and the formulation of the final product.

These biopesticides offer a safe and effective alternative to chemical pesticides and can be used in a variety of settings, including agriculture, forestry, and public health.

Challenges for Biopesticide Formulation

The biological activity or mode of action of a pesticide is determined by its biologically active microorganism or metabolite. The active ingredient in its raw form requires formulation to facilitate its handling, storage and application.

A key difference between the formulation of conventional (chemical) pesticides and the formulation of biopesticides is the living - often proteinous - nature of the biopesticide, and their biological viability is sensitive to storage conditions and environment.

A further challenge for the formulation of microbial biopesticides is the existence of additional fermentation materials besides the microbes themselves. These materials may react with formulation ingredients.

Another challenge facing microbial biopesticide formulators is the tendency for gram-positive bacteria such as Bacillus spp. to produce a slime layer relating to endospore formation.

Suspension of these bioformulations for spraying can lead to excessive sliminess and the blockage of sprayer filters and nozzles. While the slime layer is removable *e.g.* by centrifugation, the slime layer can contain components such as lipopeptides, which may contribute to the biopesticidal effect.

Shelf life (storage stability)

Other challenges to biopesticide formulation include the shorter shelf life of biopesticides compared to conventional pesticides. Significant formulation efforts are being targeted at the long-term stabilization of microbial and other biopesticides.

Shelf life measured in years is normal for conventional plant protection products, while for biopesticides a shelf life of 1 to 2 years is desirable.

For many biopesticides, however, shelf life is measured in weeks, and the correct storage of these products is vital. This is especially important for non-spore-forming, gram-negative bacteria.

Shelf life and viability may be extended by reducing the storage temperature, through freezing or refrigeration. While refrigeration is not an option in most cases, exceptions do exist for high-value crops and there appears to be a growing tendency for growers to treat biopesticides as perishable high-value products.

Microorganisms may be frozen or refrigerated in culture broth or an appropriate buffer. However, depending on the microorganism, storage in culture broth may impact long-term cell viability, and extensive testing and development in this area is a high priority.

Besides low temperature storage, longer shelf life may be obtained by increasing the number of microorganisms in the product to ensure long term viability despite a decline in their population.

Fermentation parameters such as the rate of dehydration and the physiological state of the bacteria on harvest from the medium will affect the shelf life of the final formulation.

Semiochemical formulation

Semiochemicals - natural compounds such as pheromones - are difficult to formulate due to their volatility.

Their encapsulation in traditional polymer matrixes or natural biological microcapsules such as yeast cells may prevent their dissipation and degradation by allowing their controlled release to the environment.

Microcapsules are of about the same size as pollen grains, and microencapsulated biopesticides may pose a hazard to foraging bees. These may carry the microcapsules back to their hives where the biopesticide is released, with the risk of harming the bees.

Formulation Delivery

Due to the particulate nature of most biopesticides, agitation of the spray suspension is vital to prevent precipitation in the spray tank.

Biopesticides are formulated for use with standard application equipment, and additional levels of filtering and straining as well as the use of large-orifice nozzles are recommended to prevent clogging.

Biopesticides are exposed to detrimental environmental conditions when applied in the field. These conditions include sunlight (UV), fluctuating moisture conditions, the influence of plant physiology and biochemistry as well as competition by microorganisms already present on the leaf surface. These factors may lead to shorter persistence once the biopesticide is applied to crops (field persistence).

UV-stability may be addressed by the addition of protectants (sunscreens such as oxybenzone and light blockers, such as lignin), the inclusion of antioxidant-rich natural products, and through the encapsulation of the active metabolite or microorganism.

Future developments include expressing of genes coding for enzymes capable of repairing UV-damaged DNA. In addition, an understanding of crop physiology may be used to reduce the exposure of biopesticides to sunlight, *e.g.* by utilizing the physiology of maize leaf whorls to funnel biopesticide granules to the feeding sites of larvae.

Since many biopesticides are slow to control the pest or pathogen, growers may feel that the product has insufficient effect and reapply unnecessarily.

Diligent reading of labels and understanding how the product works, as well as appropriate scouting is necessary for the optimal application of these products, allowing growers to reap the full benefits and attain consistent results.

Formulation types

Many of the formulation types described in the previous section on conventional pesticide formulation are relevant for biopesticide formulation.

Due to differences in environmental conditions, biopesticides may be formulated in different forms, each designed for a specific market.

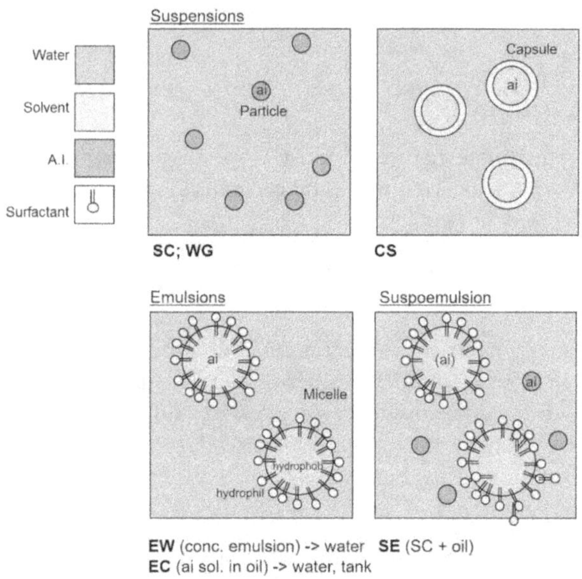

Figure 17.1: Summary of selected formulation types in aqueous spray solutions.

Most biopesticides are formulated as dry (solid) formulations. Dry formulations of dormant bacterial spores tend to have a longer shelf life compared to most liquid formulations.

Dry formulations do, however, have several disadvantages, including low field performance and the risk of nozzle blockage.

Dry biopesticide formulations include: dusts (to be applied dry), wettable powders (WP; intended to be applied as an aqueous suspension) and granules (G), which may be applied directly or mixed into water as a suspension (water dispersible granules; WG or WDG).

Liquid biopesticide formulations comprise biopesticidal metabolites or microorganisms suspended in a medium containing adjuvants, protectants and nutrients.

They may be formulated as conventional formulations such as suspension concentrates (SC), cell suspensions in waters, oils or combinations of both (emulsions; EC or EW) and microcapsules (CS).

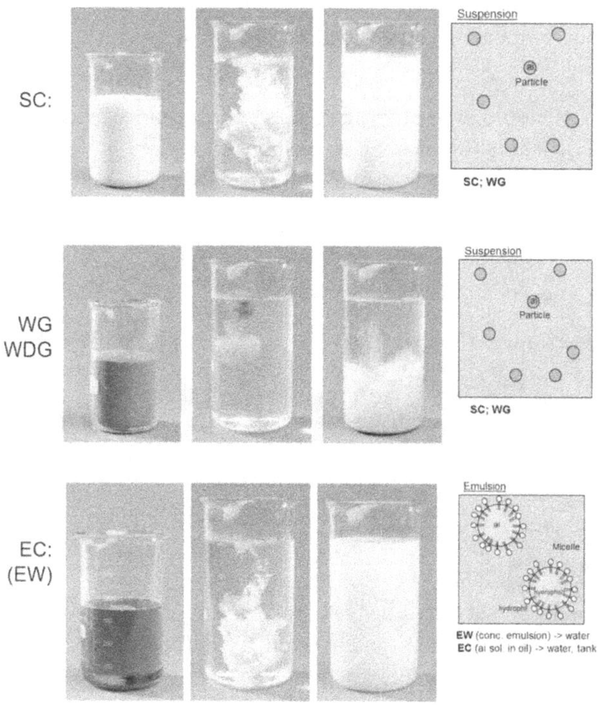

Figure 17.2: Suspensibility (SC; WG) and solubility (EC; EW) of pesticide formulations. Photos used with permission: Bayer AG, Leverkusen, Germany.

Of relevance for biopesticides, *yeast-derived microcapsules* (YC) are under development as biological delivery systems.

An added benefit of this type of formulation is that plants may recognize the yeast cells as possible invasive pathogens, leading to the induction of plant defense responses.

Liquid biopesticide formulations address many of the issues biopesticides face regarding shelf life and field efficacy, and may generally be applied using field equipment without issue.

The latter is of importance for the acceptance of biopesticides by growers, who usually are not willing to invest in new equipment specifically for the application of biopesticides.

. . .

Inerts and Adjuvants

Formulation components other than the active ingredient or microorganism are called inerts.

Inerts may contribute to improving product performance by optimizing contact to leaf, soil or insect surfaces, by reducing desiccation, by increasing shelf life or by increasing solubility or suspensibility in water.

Inerts can include dry carriers (mineral: *e.g.* clay, or organic: *e.g.* flour) or liquid carriers (*e.g.* vegetable oils) as well as binders, desiccants, thickeners, surfactants, stickers and UV protectants.

Formulation carriers can affect survival and viability, by affecting both drying-survival and storage stability.

To ensure microbial survival and efficacy after delivery, formulations may contain moisture-retaining polymers and carbon sources (nutrients) such as molasses or peptone.

Hydrogels consist of a network of hydrophilic polymer chains which can hold large amounts of water.

Hydrogels may be used in slow release biopesticide formulations to improve nutrient and moisture retention as well as reduce degradation, leading to improved biopesticide viability and efficacy.

For biopesticides, it is important that formulation inerts have low toxicity, both from an environmental and health perspective, but also with respect to the viability of the microorganism being formulated.

Chitosan is often included in biopesticide formulations (for example, in hydrogel formation), and its possible antibacterial effects must be considered when formulating microbial biopesticides.

While many conventional adjuvants are suitable for biopesticide applications, special consideration must be given for the living nature of biopesticide products.

The objectives of including *adjuvants* in biopesticide formulations are to stabilize the active metabolite or microorganism, to optimize delivery to the target, to protect the biopesticide during storage and in the field, and to enhance the efficacy of the biopesticide.

Due to the particulate nature of most biopesticides, suspension agents as well as agitation of the spray suspension is required to prevent precipitation in the spray tank.

Biopesticides are sensitive to sunlight. UV-stability may be addressed by the addition of protectant adjuvants (sunscreens and light blockers).

To prevent desiccation and ensure microbial survival and efficacy after delivery, formulations may contain moisture-retaining polymers (*e.g.* hydrogels) and carbon sources (nutrients) such as molasses or peptone.

When choosing adjuvants to optimize microbial growth, it may be necessary to consider whether they also stimulate and support pathogen germination and infection.

Finally, it is important that adjuvants for biopesticides have low toxicity, both from an environmental and health perspective, but also with respect to the viability of the microorganism being formulated.

Compatibility of Biologicals with Agrochemicals

The compatibility of biopesticides with conventional agrochemicals is an important consideration, as the two types of pest control products may be used together in some cases. In general, biopesticides are compatible with many conventional agrochemicals, but there may be some instances where the combination of the two types of products is not recommended. The following provides an overview of factors to consider, as well as options to address potential compatibility issues:

Factor	Example	Risk of mutual antagonism	Options
The living nature of the biopesticide	Synthetic ai's and inerts may be biocidal	Degradation, effects on target biology, attraction...	Compartmentalization (SE, OD, CS, nanotechnologies, sprayers)
The compatibility of pesticidally active metabolites	Metabolites and synthetics have different chemistries, sensitivities to pH, salinity etc.	Degradation, uptake inhibition, environmental response...	Compartmentalization (formulation- & nanotec, sprayers), multifunctional inerts, compatibilizers
The particulate nature of microbial biopesticides (Bacillus spp. sliminess)	Could affect nozzle and pressure choices, spray solution stability etc.	Aggregation, spray tank precipitation, nozzle blockage, sprayability, reduced bioavailability	Multi tank / nozzle spray technologies
The presence of fermentation materials	Possible issues of high-electrolyte tolerance etc.	Effects on target biology, attraction, spray solution stability...	Inerts (multifunctional, bioderived inerts, compatibilizers)
Kinetics of bioavailability and efficacy	Synthetics have faster efficacy kinetics than biologicals	Inactivation of target organisms can reduce biologicals efficacy	Controlled release technologies

Figure 17.3: Factors affecting the compatibility of biopesticides with agrochemicals.

Formulation Trends

The trend in conventional and biopesticide formulation is to move away from wettable powders and suspension concentrates to water dispersible granules, and from dusts to granules.

Controlled release formulations are being developed to optimize effect, while nanotechnology is expected to provide new types of formulations such as nanoemulsions, nanosuspensions and nanocapsule formulations.

Another area of focus is the choice of appropriate adjuvants for optimal biopesticidal activity. Here, collaboration between biologists and chemists is facilitating the development and introduction of new formulations and applications, reducing the inconsistency of field performance and encouraging growers to implement this new technology.

Attention is also being given to delivery and application practices, specifically to understanding the impact of plant defense induction kinetics on application timing and placement.

Until recently, slow progress in formulation and delivery research has been a major hurdle to biopesticide development and acceptance.

Current research efforts in bioformulation technology are focused on promoting the bioefficacy following application, especially regarding application timing and placement of the formulated product.

Formulation is one of the key factors in determining the commercial success of a biopesticide candidate.

There are significant costs involved in solving the biological issues of biopesticide formulations (including storage and field stability as well efficacy consistency), but the benefits in terms of market access and the acceptance of biopesticides as viable commercial alternatives or supplements to conventional plant protection products suffice to warrant this effort.

18

BIOLOGICALS IN CROP PROTECTION FROM A HUMAN HEALTH PERSPECTIVE

In recent years any discussion of crop protection issues has inevitably led to the question: biologicals - are they a better option than chemicals? The short answer is - yes - by almost every metric, also from a human health perspective.

However, as our understanding of the mechanisms behind biologicals continues to grow, we must continually be aware that biopesticides, despite generally having a low risk potential for human health or as an environmental hazard, are not necessarily harmless, despite being of natural origin

Biopesticides are expected to increase from about 2% of the global pesticide market in 2003 to about 8% in 2020. This growth is driven by stricter pesticide regulatory requirements, consumer awareness regarding hazards caused by chemical pesticides and the need for new and sustainable modes of action are factors driving the growth of the biopesticide market.

Generally, biopesticides have a low risk potential for human health or as an environmental hazard, but it is important to be aware that biopesticides are not necessarily harmless, despite being of natural origin.

Health Considerations: Biopesticide mode of action

BioPesticides may be grouped based on their mode of action into Bioherbicides, Bioinsecticides & Bionematicides, and Biofungicides. In the following, we will consider each of these groups from a human health perspective.

. . .

Bioherbicides

Microbial bioherbicides include bacterial and fungal endophytes (microorganisms which colonize plants without causing disease) as well as viruses.

Adverse health impacts may arise following exposure to these organisms. Research regarding health impacts has primarily focused on the potential toxicity and pathogenicity, while their potential allergenicity is less studied.

Bioinsecticides

Bioinsecticides include microorganisms (such as *B.thuringiensis*) that infect insects or nematodes, and natural compounds with insecticidal activity sourced from microorganisms, plants or animals.

Some strains of *Bacillus thuringiensis* (Bt) are specific to Lepidopteran species, while other strains may target larvae of Diptera and Hemiptera.

Reports of Bt toxicity toward non-target organisms, as well as binding of the bacterial toxins to clay soils have been the topic of debate. Most researchers agree that Bt monitoring must continue, and their use modified as appropriate.

Biochemical bioinsecticides may target specific insect neural functions, such as the uptake of ions across neural membranes. Generally, bioinsecticides have a high specificity towards insects, due to their greater affinity for insect neuroreceptors, relative to neuroreceptors in mammals.

In contrast to specific insecticides, other insecticides may target metabolic processes which are common to insects and mammals. As the AChE enzyme is similar in insects and mammals, this group of insecticides is not considered selective, and appropriate care must be exercised in their application.

Accordingly, it should be noted that not all natural products are able to be registered as biopesticides. For example, despite their natural origins, pyrethrins, spinosad and abamectin are registered as conventional (chemical) pesticides due to their neurotoxic mode of action.

Biofungicides

Microbial biofungicides may contaminate and form toxins in food materials, leading to gastrointestinal illness

Biochemical biofungicides include those able to induce endogenous plant defenses in the form of secondary plant metabolites.

While biopesticidal elicitors of plant defense responses themselves may be benign, the induced secondary plant metabolites may pose a toxicological risk.

From the point of view of potential for human health or environmental hazard, we should bear in mind that natural toxins may be many times more toxic than synthetic toxins.

A key factor driving the probable introduction of stricter EU and National regulations is thus the increasing focus on the potential health and environmental impacts of natural plant toxins or metabolites produced through the induction of plant defense responses by biopesticides.

Similarly, for organic crops which are dependent on the induction of natural plant defenses against pests and pathogens, the human and environmental health perspectives of plant-defense related toxins are the subject of growing research interest.

Biopesticide Formulation Inerts

For many biopesticides, formulation components are inert or of no toxicological concern, and risk assessment may be based on the active substance alone, and on the basis of scientific evidence.

The term "inerts" can, however, be misleading, as it implies these components do not have a particular function or that they are benign from a human and environmental health perspective.

Inerts can have health and ecosystem impacts, including endocrine disruption and allergenic effects. Thus, a biopesticide formulation may combine a benign active ingredient in a formulation that includes problematic inert ingredients.

Health considerations: the environmental fate of Biopesticides

The cited advantages of biopesticides should always be considered critically. Common benefits mentioned include:

"Biopesticides are generally biodegradable" While Biopesticides are generally rapidly degraded - a factor cited as an ecotoxicological advantage, but as a disadvantage with respect to product stability under field conditions - care should be taken to avoid excessive generalization. As many biopesticides are subject to microbial and UV degradation, the rate of degradation may significantly decrease under anaerobic conditions, and in the absence of sunlight. For example, while pyrethrins are rapidly degraded in soil by microbial activity and photolysis, they experience reduced degradation in anaerobic sediments in the absence of light and may present a toxicological risk to aquatic organisms.

"Biopesticides leave few or no harmful residues" Under EPA regulations, biopesticides may be subject to exemptions from residual tolerance levels. This is, however, contentious, as it could be argued that the statement "few or no harmful residues" is a question of definition - if the biopesticide is effective against an organism, it must per definition be harmful to that organism, and potentially to other, related or similar organisms.

"Biopesticides are less likely to harm non-target species" The OECD guidance for microbial biopesticides is that: 'the micro-organism and its metabolites pose no concerns of pathogenicity or toxicity to mammals and other non-target organisms which will likely be exposed to the microbial product; the micro-organism does not produce a known genotoxin; all additives in the microbial manufacturing product and in end-use formulations are of low toxicity and suggest little potential for human health or environmental hazard'. Biopesticides with a very specific MoA and rapidly degradation could be considered to be less likely to harm non-target species - but this applies to both biopesticides as well as conventional pesticides.

Biosafety - Biopesticide Risk Profiling and Evaluation

While microbial strains are widely prevalent in the environment, they tend to exist in equilibrium in nature which ensures a natural biosafety net.

But when specific strains are selected for their biopesticidal activity and are applied at rates sufficient to attain the desired effect levels, this may raise public health and ecological concerns.

As for conventional crop protection products, differentiation between risk and hazard is required: a Hazard is something that has the potential cause harm, while a Risk is the chance that the hazard will cause harm. Risk is often defined as Hazard x Exposure:

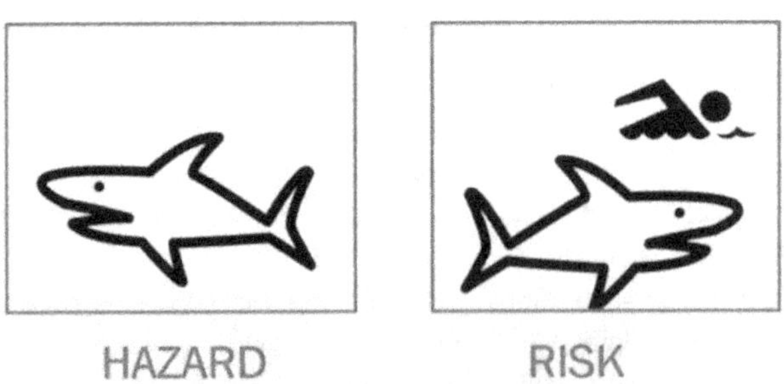

Figure 18.1: Hazard vs. Risk

It is necessary to include agrobiological considerations when evaluating microbial biopesticides for pest and pathogen control. Evaluation criteria for Risk Profiling and Evaluation include:

Route of Exposure

The common routes of exposure are oral and pulmonary uptake, ocular irritation and dermal exposure. As is the case for conventional pesticides, formulation type and method of application define the probable routes of exposure, as well as the populations exposed.

A product applied as a spray or as a dust or powder may be inhaled during handling and application or lead to eye irritancy, whereas a paste is most likely to pose a risk of dermal exposure.

The life-cycle and environmental fate of the organism may also influence the route of exposure. For example, a microbial biofungicide may provide an exposure risk to pesticide applicators, but if sporulation is expected after application, inhalation of the spores may become a significant route of exposure to a wider non-target population, including humans.

An additional risk factor for microbial biopesticides is the potential for allergic sensitization (hypersensitivity) to handlers, applicators and bystanders following exposure.

Level of Exposure

The environmental fate and life cycle of microbial biopesticides defines the probable levels of exposure.

Microbial biopesticides are generally living organisms and may have the potential to persist and replicate, as well as to spread from the site of application, affecting its relative risk potential for human health or as an environmental hazard.

To attain desired levels of efficacy, biochemical biopesticides may need to be applied at higher doses than conventional pesticides - a factor which must be considered when evaluating biopesticidal toxicity to non-target species.

Risk from Exposure

In addition to the life-cycle and environmental fate of the biopesticide, the biopesticidal mode of action, target pest biology, and ability to induce plant defense toxins determine the potential for toxic or pathogenic effects in humans and other non-target organisms.

Accordingly, a highly specific, biochemical bioherbicide might be considered less likely to cause adverse effects in humans than for example a microbial bioinsecticide with a broad animal host range and targeting a mechanism common to the physiology of the target pest as well as humans.

For living microbial organisms, the physiological ability to survive and replicate in the environment is determined by a number of factors including temperature and metabolic requirements, which define their viability under field conditions and thus their risk potential for human health or as an environmental hazard.

Conclusions

Biologics in Crop Protection are a better option than chemicals from a human health perspective.

However, as our understanding of the mechanisms behind biologicals continues to grow, we must continually be aware that biopesticides, despite generally having a low risk potential for human health or as an environmental hazard, are not necessarily harmless, despite being of natural origin.

The potential for adverse effects on human health or as environmental hazards may derive from:

- mode of action (e.g., mitochondrial inhibition as well as neurotoxic effects).
- the potential for allergic sensitization.
- secondary metabolites (e.g., induced plant defense toxins).
- formulation inerts.

Biopesticide risk profiling and evaluation may be determined by :

- *Route of exposure* - the nature of the formulation and its potential uptake routes.
- *Level of exposure* - the physiology and viability of microbial organisms in the environment and the application rates of biochemical biopesticides.
- *Risk from exposure* - the specificity of the mode of action, target host and mechanism range and climatic as well as metabolic conditions.

19

BIOPESTICIDE REGULATION

Biopesticides (microbial pesticides, bio-derived chemicals and plant-incorporated protectants) introduce unique and complex modes of action and may be used in plant production on an equal footing with chemical pesticides

Even though biopesticides have been commercially used for over 60 years, they have generally been evaluated and registered following the model for the registration of conventional pesticides, and laws and policies regulating their use vary from country to country.

Submission procedures have often been lengthy, and the high cost of registration has limited the commercialization of new and safer biopesticidal products, particularly in the EU.

However, many of the data requirements (such as chemical identity) and evaluation criteria may not be relevant for most biopesticides, and risk levels for environmental or human health safety are (generally) considered to be lower than for conventional pesticides.

In the following, an overview of biopesticide evaluation and registration is presented, with a primary focus on the regulation of biopesticides in the US and EU.

Biopesticide Regulation in the US

In the United States, the Environmental Protection Agency (EPA) is responsible for regulating pesticides, ensuring they do not pose adverse effects to humans or the environment.

The Biopesticides and Pollution Prevention Division (BPPD) falls under the EPA and is responsible for all regulatory activities associated with biopesticides.

In the US, biopesticides are recognized as a regulatory category and are regulated following a different set of data requirements than for conventional chemicals.

Biopesticide data requirements for registration are listed in 40 CFR Part 158, Data Requirements for Pesticides.

Biopesticides as a category are typically considered by EPA to be "reduced risk pesticides" - of lower risk than conventional pesticides. However, biopesticides are not necessarily harmless, despite being of natural origin - the formation of plant-produced toxins induced by biopesticides is a case in point.

Genetically modified microbial pesticides (plant-incorporated protectants) are evaluated on a case-by-case basis depending on the microorganism, proposed use pattern and nature of the genetic modification.

EPA registration is not required for semiochemicals (*e.g.* pheromones) used in fixed-location lures to attract and monitor pests, with minimal impact on either the environment or human health.

The EPA generally requires fewer data for biopesticides than for conventional pesticides (with e.g. most pheromones being exempted for testing) and biopesticides may be subject to exemptions from residual tolerance levels.

This last point, however, is contentious, as it potentially hinders global trade to regions with maximum residue levels in food (such as the EU), reducing the potential economic viability of new biopesticides and thus their introduction into global crop protection strategies.

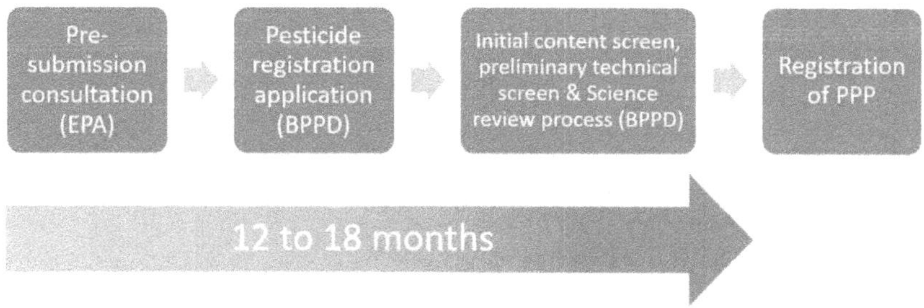

Figure 19.1: Assessment and approval timelines for biopesticide approval in the US.

The biopesticide registration process may begin with an electronic pre-submission consultation request, at which the EPA will provide specific guidance for application.

The registration process continues with the formal submission to the BPPD of a pesticide registration application, which leads to an initial content screen, followed by a preliminary technical screen.

If no issues are found, the formal science (including chemical-physical, toxicological and ecotoxicological) review process is initiated.

Registration of biopesticides takes 12 to 18 months, compared to approximately 36 months for conventional pesticides, and registration fees are lower.

Not all natural products are able to be registered as biopesticides. For example, pyrethrins, spinosad and abamectin are registered by the EPA as conventional (chemical) pesticides due to their neurotoxic mode of action.

In contrast, Azadirachtin (a botanical insecticide extracted from natural neem oil) acts as an insect growth regulator, anti-feedant, repellent, sterilant, and oviposition inhibitor and is registered as a biopesticide.

Biopesticide Regulation in the EU

Fewer biopesticides are registered in the EU than for example in the United States, South America, India or China, due to lengthy and complex registration processes in the EU, which follow a risk-based approach to the registration of pesticides – including biopesticides.

Biopesticides are regulated as plant protection products under EU plant protection Regulation 1107/2009, which - unlike the EPA - does not recognize "biopesticides" as a regulatory category.

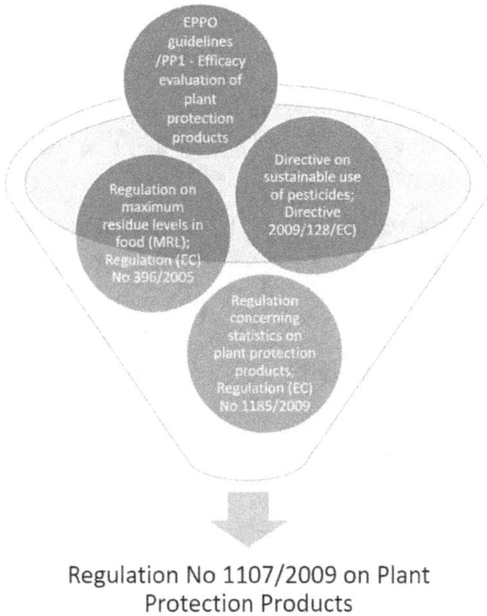

Figure 19.2: Regulation of Plant Protection Products in the EU.

Regulation is carried out in conjunction with other EU Regulations and Directives (*e.g.* the Regulation on Maximum Residue levels (MRLs) in food; Regulation 396/2005) and the Directive on sustainable use of pesticides; Directive 2009/128/EC).

Directive 2009/128/EC aims to advance the sustainable use of pesticides, Integrated Pest Management (IPM) and the use of non-chemical alternatives to pesticides - including the development and use of biological pesticides.

Following approval, a monitoring program ensures pesticide residues (Maximum Residue Levels; MRLs) in food are below the limits set by the European Food Safety Authority (EFSA).

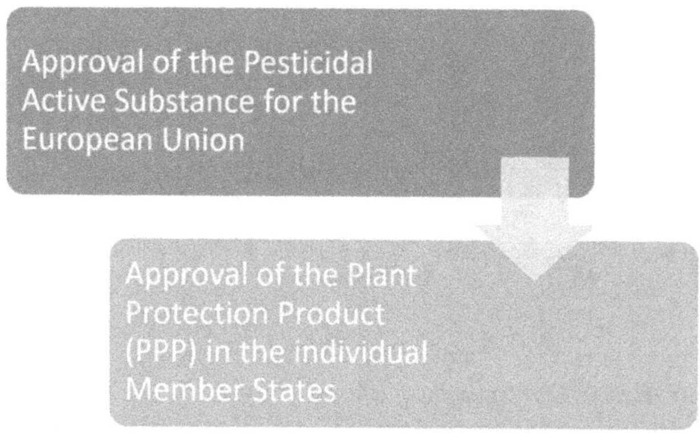

Figure 19.3: Two-tiered approach for the EU approval and authorization of pesticides.

Before pesticides can be introduced to the European market, the pesticidal active ingredient (active substance) must be approved for the European Union, after which approval of the Plant Protection Product (PPP) can be sought in individual Member States.

For the approval of a pesticidal active ingredient (active substance), the applicant submits a dossier containing supporting scientific data (including efficacy, chemical-physical, toxicological and ecotoxicological, analytical methods to determine purity and potential residues, as well as environmental fate, residues and operator exposure studies) to a Member State (Rapporteur Member State).

The Rapporteur Member State evaluates the application and, if the dossier is considered admissible, the Rapporteur Member State starts evaluating the active substance.

The Rapporteur Member State produces a Draft Assessment Report, which is submitted to the European Commission and EFSA, who is responsible for the peer reviewing of the risk assessments on active substances.

Finally, the European Commission presents a review report to the Standing Committee for Food Chain and Animal Health (SCFCAH), who votes on approval or non-approval of the active substance.

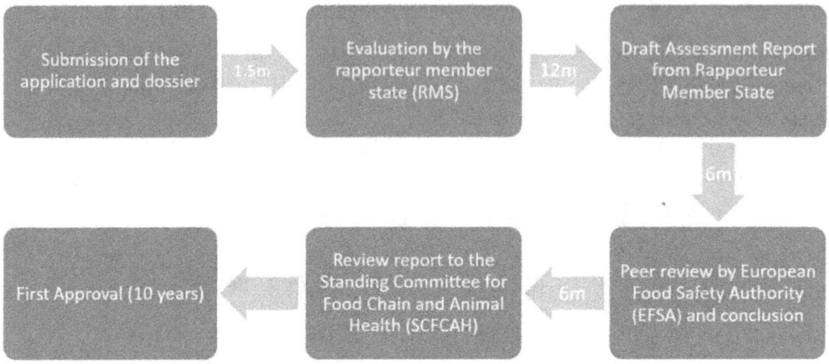

Figure 19.4: Procedure for Active Substance approval; EU (minimum duration in months, m).

The subsequent authorization of a plant protection product (PPP) is carried out by the Member States, with the applicant specifying the Member State carrying out the evaluation. Multiple dossiers (for the product, for each active ingredient, a draft product label etc.) are submitted.

Depending on the complexity and completeness of the dossier, it takes 2.5 to 3.5 years to approve a new active substance in the EU.

The authorization process for a PPP may not exceed one year. Following authorization, a company may apply for mutual recognition within one of the three geographic zones (North, Central, South) for products with the same use(s) under similar agricultural conditions.

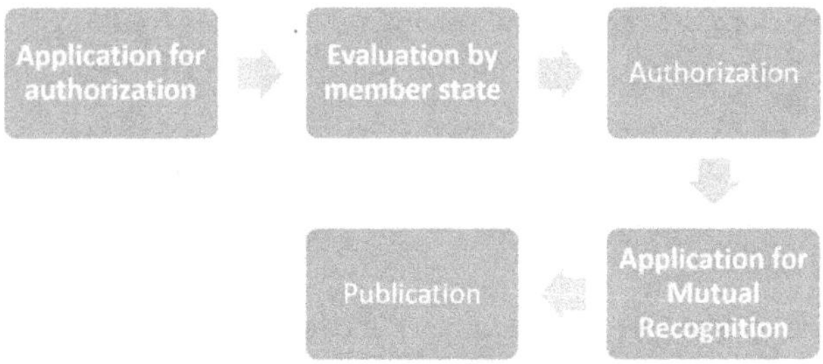

Figure 19.5: Procedure for Plant Protection Product approval; EU.

Although "biopesticides" do not exist as a regulatory category, the pesticide categories "basic substances" and "low risk substances" were introduced in

August 2017, as defined in Regulation 2017/1432, amending Regulation 1107/2009.

A risk-based approval approach is used in the EU, as biological substances are not necessarily harmless, despite being of natural origin. While a reduced dossier may be submitted for low-risk substances, this does not preclude the demonstration of sufficient efficacy.

Biopesticides should generally qualify as low-risk active substances, facilitating the faster introduction of biopesticide candidates which fall under these categories. Currently, 18 basic substances and 10 low risk substances are approved under Regulation 1107/2009.

Basic substances

Basic substances are substances not primarily intended as crop protection products, but which could provide crop protection capabilities, and their approval is based on existing evaluations carried out in accordance with other EU legislations. Authorization is issued for the entire EU for an indefinite period.

Low risk substances

Low risk substances must fulfill criteria listed in Annex II of Regulation 1107/2009. In contrast to conventional pesticide active ingredients, approval of low risk substances may be granted in part based on literature data and reasoned scientific opinions.

Microbial pesticides, baculoviruses and semiochemicals (*e.g.* pheromones) are assessed for compliance with low-risk criteria.

Assessment (evaluation) of low risk active ingredients (regardless whether of biological or synthetic origin) takes 120 days, compared to 12 months for conventional active ingredients, and low risk active ingredients are approved for 15 years, instead of the normal 10 years.

In addition, authorization fees are lower - factors promoting the introduction of low-risk products (including biopesticides) by smaller companies.

	Basic Substances	Low Risk Active Substances
1	Chitosan hydrochloride	COS-OGA
2	Equisetum arvense L.	Cerevisane
3	Sucrose	Pepino mosaic virus strain CH2 isolate 1906
4	Calcium hydroxide	Ferric phosphate
5	Lecithins	Isaria fumosorosea Apopka strain 97
6	Salix spp. cortex	Saccharomyces cerevisiae strain LAS02
7	Vinegar	Trichoderma atroviride strain SC1
8	Fructose	Mild Pepino Mosaic Virus isolate VC 1
9	Sodium hydrogen carbonate	Mild Pepino Mosaic Virus isolate VX 1
10	Diammonium phosphate	Bacillus amyloliquefaciens strain FZB24
11	Whey	
12	Sunflower oil	
13	Hydrogen peroxide	
14	Urtica spp.	
15	Clayed charcoal	
16	Sodium chloride	
17	Mustard seeds powder	
18	Beer	

Figure 19.6: Low Risk Active Substances and Basic Substances approved under Reg. (EC) No 1107/2009.

For biopesticide development projects, due diligence should be applied to study design to ensure that the scientific requirements for potential inclusion in a registration dossier are met.

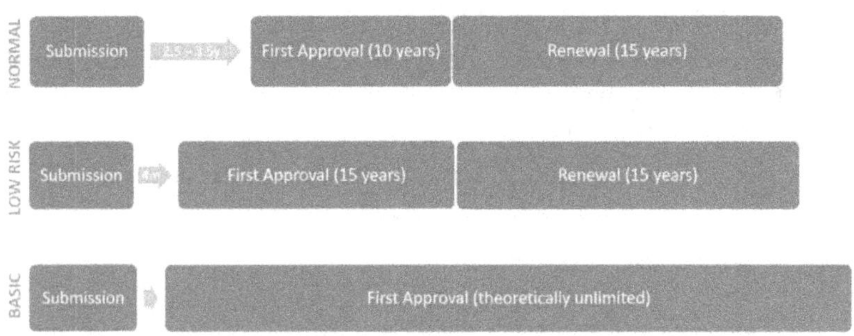

Figure 19.7: Assessment and approval timelines (y/m; years / months) for active substance approval; EU special cases.

Many biopesticidal active substances (both bio-derived chemicals and microbes) are well-documented, and significant parts of a dossier could comprise information from published literature.

Waivers for data on residues, environmental fate and ecotoxicology could be considered based on scientific evidence, keeping in mind that natural substances are not necessarily harmless.

For groups of microorganisms having common properties, exchangeability of data could reduce registration time and facilitate the rapid introduction of new technologies for biopesticide strategies.

For many biopesticides, formulation components are inert or of no toxicological concern, and risk assessment could be based on the active substance alone, and on the basis of scientific evidence.

Harmonization of Registration Guidelines

As a result, registration guidance for biopesticides is being amended at international, national and regional levels by several organizations, including FAO, the International Organization for Biological Control (IOBC), European and Mediterranean Plant Protection Organization (EPPO), the Organization for Economic and Co-operative Development (OECD), the European Union and the United States Environmental Protection Agency (EPA).

Many (but not all) countries and organizations differentiate between biopesticide and conventional pesticide registration, and advocate a harmonized approach to biopesticide registration, either through provisions for biopesticides in existing pesticide legislation, or through separate legislation for biopesticides.

Harmonization permits the exchangeability of data across countries and regulatory agencies, and regulations promoting the registration of low-risk compounds will facilitate the commercialization and use of biopesticides in crop protection strategies.

Further optimization

At global and regional levels, harmonization of accepted residue levels (tolerance levels) could promote global trade, increasing the economic viability and introduction of new biopesticide strategies.

While Regulation 2017/1432 (introducing the pesticide categories "basic substances" and "low risk substances") facilitates the development and introduction of sustainable, non-chemical alternatives to pesticides in accordance with directive 2009/128/EC, further optimization of the regulatory process could advance the use of biopesticides in the EU.

Within the EU, harmonization of the implementation of Regulation 1107/2009 between Member States could accelerate regional acceptance of biopesticidal strategies.

Although "fast-track" approval of low risk active ingredients is beneficial, the registration process could be further optimized through a staged approval process: an initial rapid assessment of low risk criteria during the development phase to be followed by full evaluation.

Finally, certain plant biostimulants - currently regulated as fertilizers - could be considered for inclusion in a “special cases” category of the plant protection regulatory framework, exploiting their benefits to crop vitality as a factor in pest and pathogen resistance.

ESSENTIAL BIOSTATISTICS

FOR PESTICIDE & BIOPESTICIDE R&D

I AM NOT A STATISTICIAN!

The information presented here comprises approximations and rules of thumb which have proven useful in designing and analyzing non-critical trials.

For critical trials, ALWAYS seek the advice of a qualified statistician.

INTRODUCTION

During the more than 20 years I have spent in Crop Protection research and research management, I came to realize that while I and many of my colleagues had followed the same courses in biostatistics, very few of us could apply the methods we had learnt, or analyze and understand the data generated, without repeatedly referring to our textbooks.

In addition, it soon became apparent that very few of the methods we had learnt were sufficient to analyze the majority of our trials.

To address these issues, I prepared a series of internal courses in these most-used "quick-and-dirty" techniques, approximations and rules-of-thumb for my laboratory colleagues, and these courses proved so popular I was asked to compile them as a printed manual.

The result is presented in this book - aimed at scientists and research workers not necessarily trained in mathematical theory, but who have previously followed a course in Biological Statistics, and who need a readily accessible overview on how to plan, implement and analyze "day-to-day" experiments without access to a dedicated staff of statisticians.

It contains few calculations (the "how" of biostatistics) but instead provides an overview of the "why" - what is it the numbers are telling us, and how can we use this to plan trials, understand our data and make decisions. It is designed to fit in a lab coat pocket for easy access to enough information to allow us to be better prepared than 90% (±5%) of our colleagues!

INTRODUCTION

As my background is in plant biology and crop protection, I use examples from the agricultural sciences. The biostatistical tests, approximations and rules of thumb described are however applicable to many other branches of biological experimentation.

A SHORT HISTORY OF APPLIED BIOSTATISTICS

The first quarter of the 20th century was an intense period of statistical development driven in part by advances in the science of genetics, and in part by the need to develop mathematical methods for the study of heredity and evolution.

It culminated by providing non-mathematical scientists with computationally simple tools for the design and analysis of their experiments based on elementary algebra, rather than on complex matrix functions and integrals.

Most of the statistical methods used by researchers in the biological sciences today stem from the insight of a handful of statisticians, and their interactions.

Four statistical pioneers in particular are immortalized in the distributions and statistical methodologies named after them.

Their influence is so significant to contemporary research in the biological sciences that the history and origins of their insights, statistical distributions and tests - in such common use today - is worth recounting.

English mathematician and biostatistician **Karl Pearson (1857-1936)** is credited with the establishment of modern mathematical statistics.

Among many other contributions to statistics, Karl Pearson introduced the p-value (Chapter 4), or probability value - the probability of finding the observed results when the null hypothesis is true as well as the term "standard deviation" (Chapter 5) in 1894 to replace the earlier "root mean square error" and "error of mean square".

In addition, Karl Pearson created the first volume of statistical tables in "Tables for statisticians and biometricians" (1914).

George Waddel Snedecor (1881-1974) was an American mathematician and statistician, and a pioneer of modern applied (as opposed to mathematical) statistics. Snedecor was dedicated to developing statistical methods for scientists not trained in mathematical theory, facilitating the interpretation of data within the biological sciences.

Snedecor's "Statistical methods applied to experiments in agriculture and biology" (1937) and "Statistical methods" (1938) are considered seminal in the development of applied statistics for research workers.

Perhaps the greatest name in the developing field of applied statistics for the biological sciences is that of British statistician and geneticist **Ronald A. Fisher (1892-1962)**.

Fisher contributed to the design and statistical analysis of experiments for non-mathematical research workers through his position as a statistician at the Rothamsted Agricultural Experiment Station, the oldest agricultural research institute in the United Kingdom.

Fisher's most important contributions to statistical methods include the introduction of the concept of randomization (Chapter 8), and the analysis of variance (ANOVA; Chapter 18) to compare the means of multiple samples, providing solutions to issues of variance arising from factors such as the heterogeneity of soils and the variability of biological material.

Several of his most important contributions were published in "Statistical methods for research workers" (1925).

Fisher focused on the development of methods for small samples (typical for the crop studies carried out at Rothamsted) and identified the distributions of many sample statistics, publishing "The design of experiments" in 1935 and "Statistical tables" in 1947 (following on from Karl Pearson's previously published book of tables.

William Sealy Gosset (1876-1937) was a chemist and statistician hired by the Guinness Brewery to apply biochemistry and statistics to Guinness' industrial processes.

Gosset introduced the z-statistic to test the means of small, normally-distributed samples for quality control in brewing, and published it anonymously under the name "Student" in Biometrika (1908), following a sojourn at Karl Pearson's laboratory in London.

Gosset's test was adapted by Fisher to incorporate the concept of "degrees of freedom" (introduced by him in its statistical context) in a 1924 paper.

Gosset himself appears to have introduced the t-nomenclature (t-distribution; t-test) in correspondence with Fisher. The t-statistic (Chapter 15) is thus also termed Students t-distribution and Students t-test.

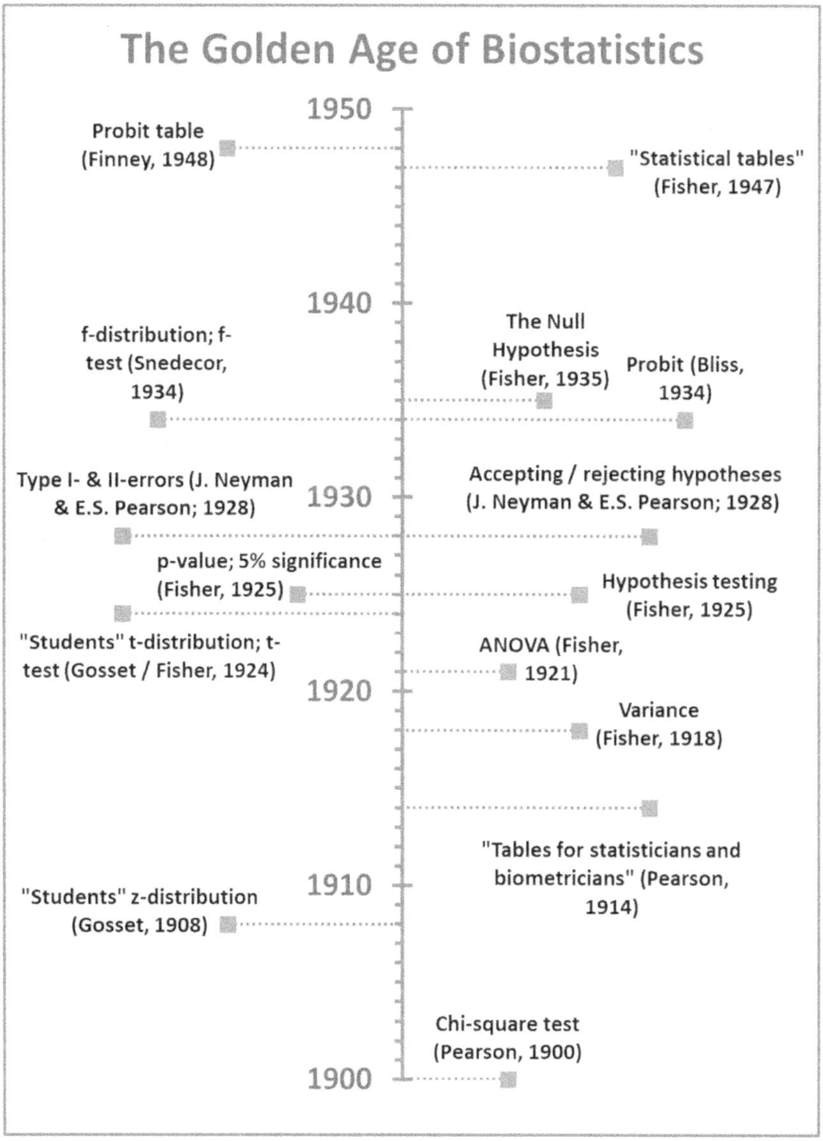

In contrast to Pearson (who sought correlations in large samples), Fisher and Gosset used small samples more typical for agricultural trials and sought causes rather than correlations.

Their differences in statistical approaches lead to increasingly bitter disputes between Pearson and Fisher, leading Fisher to decline the post of chief statistician under Karl Pearson and fortuitously (especially for experimenters in the biological sciences) chose the position as a statistician at Rothamsted instead.

In contrast, Gosset was able to maintain good relations with both Pearson and Fisher - an admirable feat of diplomacy possibly nurtured by his passion for beer... Gosset would go on to become Chief Brewer at Guinness.

Of relevance for the contents of this book, hypothesis testing in its modern form was extensively applied in R.A. Fishers "Statistical methods for research workers" (1925), while the concept of accepting or rejecting hypotheses (Chapter 1) and of Type-I and Type-II error (Chapter 3); originally referred to as "the first source of error" and "the second source of error") were presented by J. Neyman and E.S. Pearson (son of Karl Pearson) in 1928.

Fisher introduced the null hypothesis (Chapter 2) in "The design of experiments" (1935), noting that the normal hypothesis is never proved or established but is possibly disproved in the course of experimentation".

Although the p-value was introduced by Karl Pearson, it was extensively applied and disseminated in R.A. Fishers "Statistical methods for research workers" (1925).

Here, he advocated the 5% significance level (Chapter 4), identifying that 1.96 standard deviations around the mean is the approximate value of the 97.5 percentile point of the normal distribution.

This in turn led to the rule of thumb of two standard deviations for statistical significance in a normal distribution. One cannot help noticing the irony in light of the ongoing (often heated) debate regarding the use of p-values rather than fixed significance levels!

The f-test (Chapter 15) was initially developed by Fisher as the variance ratio (defined as the ratio of explained variance (or between-group variability) to unexplained variance (or within-group variability) in the 1920s.

The f-distribution was subsequently tabulated, and the f-test named, by Snedecor in 1934 as an improved presentation of Fisher's Analysis of Variance, in order to facilitate its interpretation within the biological sciences.

Probit ("probability unit") models (Chapter 19) were developed by American biologist and statistician **Chester Bliss (1899 - 1979)** in 1934, as a convenient method to evaluate dose response in pesticide data.

Probit allowed researchers to convert mortality (effect) percentages to probit values, which approximated a straight line function between the logarithm of the dose and effect which can be analyzed by simple linear regression methods

The Probit model was further adapted and tabulated at Rothamsted by British statisticians **D. J. Finney (1917 - 2018)** and W. L. Stevens in 1948 to avoid having to work with negative probits in an era before the ready availability of electronic computing.

It is these Probit tables that even today ensure that dose-response fitting to evaluate dose-response relationships may be conveniently performed when statistical software packages are not available, and experimenters do not have a background in mathematics.

It is unlikely that we will again experience a period as dynamic as that of the early 20th century for the advancement of statistical methods in the biological sciences.

We should continue to recognize the contributions of the period's leading statisticians in developing experimental planning and analysis methods capable of driving the advances in Science of the 20th century - methods which are still applicable and in use almost a century later.

THE SCIENTIFIC METHOD

The Scientific Method is the mainstay of R&D and provides an objective and systematic approach to experimentation, by minimizing the influence of experimenter bias or prejudice in the form of preconceived notions.

By adhering to the rigorous methodology of the scientific method, experimenters identify and account for errors arising from experimenter bias and fallacies. Bias may be defined as a systematic error affecting objectivity, distorting experimental findings by selecting one outcome above another.

Confirmation bias is an example in which the tendency is to focus on data supporting a hypothesis while suppressing or ignoring data that does not. Other biases include sampling bias (omitting or including groups from the sample), measurement bias (errors in data measurement) and reporting bias (the tendency to report positive results).

In its simplest form the Scientific Method can be broken down into seven phases:

1. OBSERVE. The observation of an interesting phenomenon.

2. QUESTION. The observed phenomenon gives rise to a question.

3. HYPOTHESIZE. An explanatory hypothesis is proposed to explain the phenomenon, conducting background research if necessary.

Figure 1: Summary of the Scientific Method.

4. EXPERIMENT. Designing and implementing one or more experiments to test the hypothesis. Experiments should be designed to minimize possible errors and bias using appropriate scientific controls and randomization.

5. ANALYZE. Recording observations and analyzing the data. Statistical Trial Planning and Data Analysis are key components of the Scientific Method.

6. EVALUATE. Interpret the data, deciding whether to accept or reject the hypothesis. Publish the results and draw conclusions that serve as a starting point for new hypotheses.

7. ITERATE. Running new experiments to test alternative hypotheses if the original hypothesis is not accepted, or iterative experiments to confirm and/or expand our knowledge if the original hypothesis is accepted.

Scientific method example: smartphone charging.

To obtain a more intuitive understanding, let us apply the Scientific Method to an everyday situation - a smartphone that is not charging.

In this process, we will minimize the influence of experimenter bias or prejudice in the form of preconceived notions.

1. Observe

The observation of an interesting phenomenon: Your smartphone has been on the charger all night, but this morning your smartphone was not charged.

2. Question

The observed phenomenon gives rise to a question: Why has my smartphone not charged?

3. Hypothesize

We can hypothesize that IF the smartphone didn't charge THEN the charger is defective. This hypothesis may then be reworded as a *research* or *working* hypothesis.

Accordingly, you could borrow a charger and see if your smartphone charges - you plug in the replacement charger and attach your smartphone. As we will see later, an experiment has an independent as well as a dependent variable.

The *independent variable* (the charger) can be changed or controlled, and the effect it has on the *dependent variable* (the smartphone) can be recorded and analyzed.

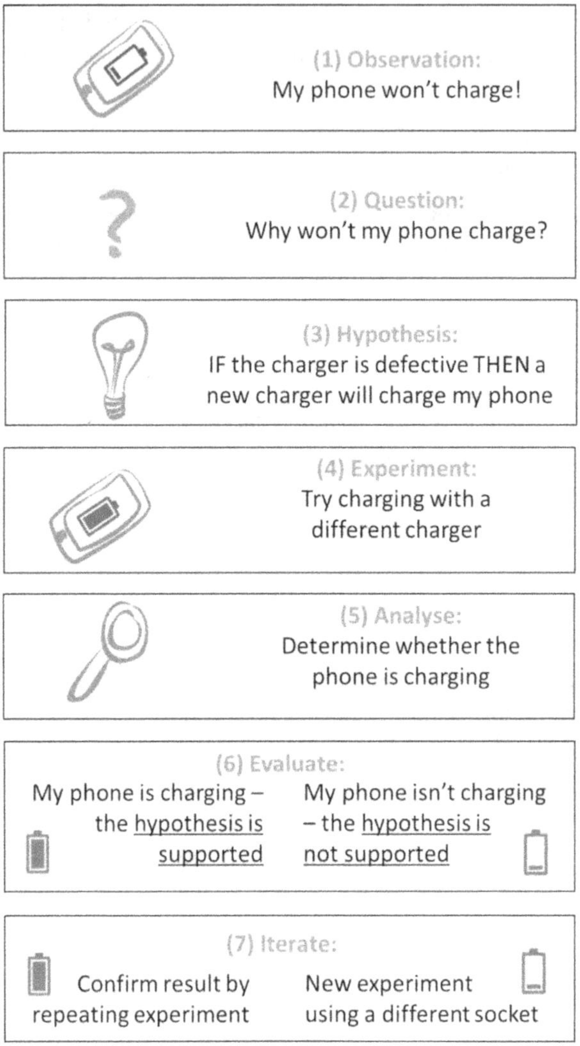

Figure 2: The Scientific Method - smartphone charging.

It is important to change only one variable for an experiment - for example: if we designed an experiment comprising both a replacement charger as well as a replacement smartphone, we would not be able to deduce anything from the data – we would not know if the issue arose from the original charger or smartphone. From this example, we can intuitively understand the importance of a correctly formulated hypothesis.

• • •

5. Analyze

You observe and note whether your smartphone begins to charge on the replacement charger. Ideally, you would repeat this several times to be quite sure of your observations.

In our example, we might repeat the charging process ten times and discover that the replacement charger only charged in eight of these instances.

Don't throw out the observations you think may be wrong - these provide valuable information which might suggest that it is not the charger that is defective but instead may be the socket which is providing intermittent power.

6. Evaluate

If your smartphone charges consistently, then the hypothesis may be considered to be supported. Alternatively, if your smartphone does not charge consistently, then the hypothesis may be rejected.

In practice, however, we can never be sure that the charger functions consistently (for example the socket may only have a 1% misfunction rate, which we would not observe with few replicates).

Thus, although the evidence supports the hypothesis, it does not necessarily mean that the hypothesis is correct. We will consider the implications of this in more detail in the next section.

7. Iterate

If your smartphone charges on the replacement charger, supporting the hypothesis that the original charger is defective, you might propose a new, testable hypothesis: IF the charger is defective THEN the charging tip may need replacing.

Alternatively, if your smartphone does not charge on the replacement charger, you might propose an alternative hypothesis: IF the charger is not effective THEN the power socket may be defective.

The above is a simple example intended to illustrate the steps of the scientific method and provide an intuitive understanding. In R&D the process is far more complicated and requires us to dig a little deeper into the underlying theory.

With this information on the Scientific Method, we are ready to move on to the next section: Proposing a Hypothesis.

I

PROPOSING A HYPOTHESIS

1

HYPOTHESES

Several years ago, I was part of a research team working on the development of herbicide safeners.

> Herbicide safeners are (typically) chemicals which protect crops from herbicide injury by accelerating the metabolism of herbicides, or by inhibiting their translocation within plants, facilitating selectivity between crop plants and weed species targeted by the herbicide.

At the time we ran herbicide safener experiments in a dedicated section of the greenhouse which was adjacent to a section containing plants under evaluation for leaf damage (phytotoxicity) as a result of treatment with one of our commercial insecticide products.

We had hit a wall with regard to identifying a formulation additive which could reduce phytotoxicity without reducing insecticidal activity. The project had been placed on hold while we considered our next steps.

Shortly after starting the herbicides safener experiments, we observed that herbicide-treated plants showed the same symptoms of oxidative damage as the adjacent insecticide-treated plants. An idea began to germinate...

According to the precepts of the scientific method, we had made an interesting observation: the symptoms of insecticide phytotoxicity appeared identical to those of oxidative herbicide phytotoxicity.

A quick literature study confirmed that leaf damage by insecticides (phytotoxicity) could in many cases be ascribed to membrane peroxidation arising from reactive oxygen species.

The herbicide safener being tested reduced herbicidal symptoms so effectively that we questioned whether it would be able to reduce the phytotoxicity of the insecticide - perhaps this was the elusive formulation additive which could reduce phytotoxicity without reducing insecticidal activity? We were on our way to developing a hypothesis!

A hypothesis is the basic tenet of the scientific method. It is often described as an "educated guess" to account for an observed phenomenon or a research idea and is based on prior knowledge and observation[1].

A *Research* or *Working* Hypothesis is a proposal or prediction to be proved or disproved through experimentation or observation.

A Research Hypothesis may be formulated using Boolean IF / THEN logic:

IF the charger is defective, THEN a new charger will charge my phone.

IF herbicide safeners prevent herbicide phytotoxicity THEN herbicide safeners could prevent insecticide phytotoxicity.

In the previous section, it was determined that if your smartphone charges consistently, then the hypothesis may be considered to be supported.

In practice, however, we can never be sure that the charger functions consistently (for example the socket may have a 1% malfunction rate, which we would not observe with a few replicates). Thus, although the evidence *supports* the hypothesis, it does not necessarily *prove* that the hypothesis is correct.

A fundamental aspect of the scientific method is Karl Popper's[2] Principle of Falsification: for any hypothesis to be valid, it must be disprovable before it can become accepted as a scientific hypothesis.

Put simply: repeated observations of white swans do not prove the hypothesis that all swans are white, but the observation of a single black swan[3] disproves the hypothesis.

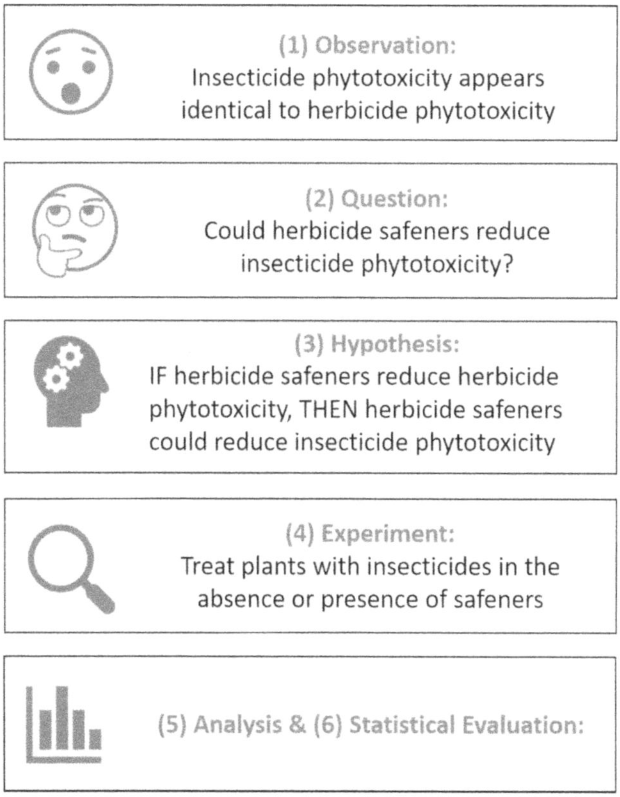

Figure 1.1: Applying the Scientific Method.

Without deviating further down this philosophical rabbit hole, we say that rather than *proving* the working hypothesis, researchers follow the Popperian Scientific method where 'nothing in the sciences can be proven, only disproven'.

Researchers thus work to reject or *disprove* the *Null hypothesis* - the default hypothesis that 'nothing happened'. Researchers also devise an *alternate* hypothesis - one that they think explains a phenomenon - but work to reject the Null hypothesis.

---

1. Hypothesis testing in its modern form was extensively applied in R.A. Fishers "Statistical methods for research workers" (1925), while the concept of accepting or rejecting hypotheses was introduced by J. Neyman and E.S. Pearson (son of Karl Pearson) in 1928.

2. Karl Popper (1902-1994): influential science philosopher, who introduced the concept that science advances by "deductive falsification" *i.e.* experiments *test* theories rather than produce them.
3. Until 1697 when Dutch explorers became the first Europeans to see black swans (*Cygnus atratus*) in Australia, it was presumed - at least by European scientists - that black swans did not exist.

2

THE NULL HYPOTHESIS

The *Null hypothesis* **(H0)**, is the hypothesis that the data generated by an experiment occurred by mere chance and was not affected by the treatment - "nothing happened".

The *alternative hypothesis* **(H1)** is the hypothesis that the treatment influenced the data generated, "something *did* happen".

So, for our herbicides safener case an appropriate research question was: *could herbicide safeners effectively reduce insecticide phytotoxicity?*

From this a Research hypothesis could be formulated: *"IF herbicide safeners reduce herbicide phytotoxicity, THEN herbicide safeners could reduce insecticide phytotoxicity"*, and the associated Null hypothesis and its alternate may take the form:

Null hypothesis (H0): There is no difference between treatments.

Alternative hypothesis (H1): There is a difference between treatments.

Following experimentation and the statistical evaluation of whether the Null hypothesis provides a valid explanation of the data, the Null hypothesis may be rejected or accepted *with a certain level of probability, or confidence* (for example, 95% confidence - more on this later).

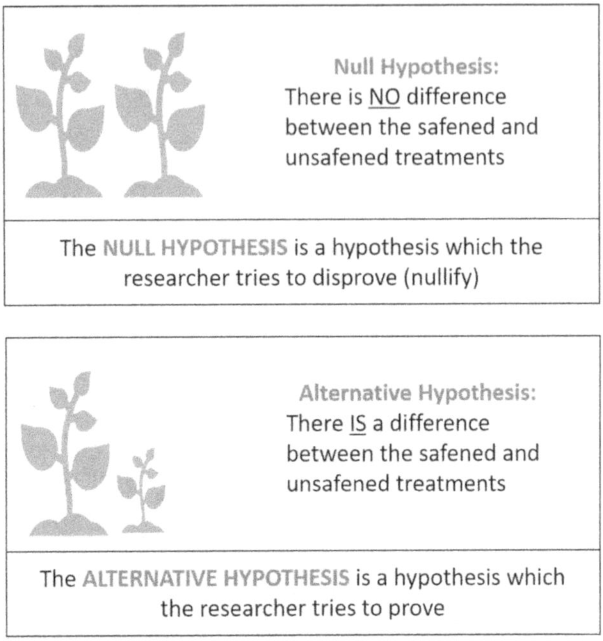

Figure 2.1: The Null Hypothesis.

This element of probability arises because we are generally not able to test the entire population and work with samples instead, which brings in a "black swan" element of probability.

Thus, we speak of "rejecting the Null hypothesis in favor of the alternative hypothesis, and "failing to reject the Null hypothesis" rather than of "accepting the Null hypothesis", as we have no way of proving that the data occurred by chance[1].

Similarly, *"failed to prove the prisoner is guilty"* is not the same as *"proved the prisoner is innocent"*.

In the example below, if there is a difference between the safened and unsafened treatments, we reject the Null hypothesis that "nothing happened" and shift our belief to the alternate hypothesis that "something *did* happen" (although the latter has not been proven).

With this information on the Scientific Method and Hypotheses, we are ready to move on to the next section: Designing and implementing experiments (Significance, Power and Effect).

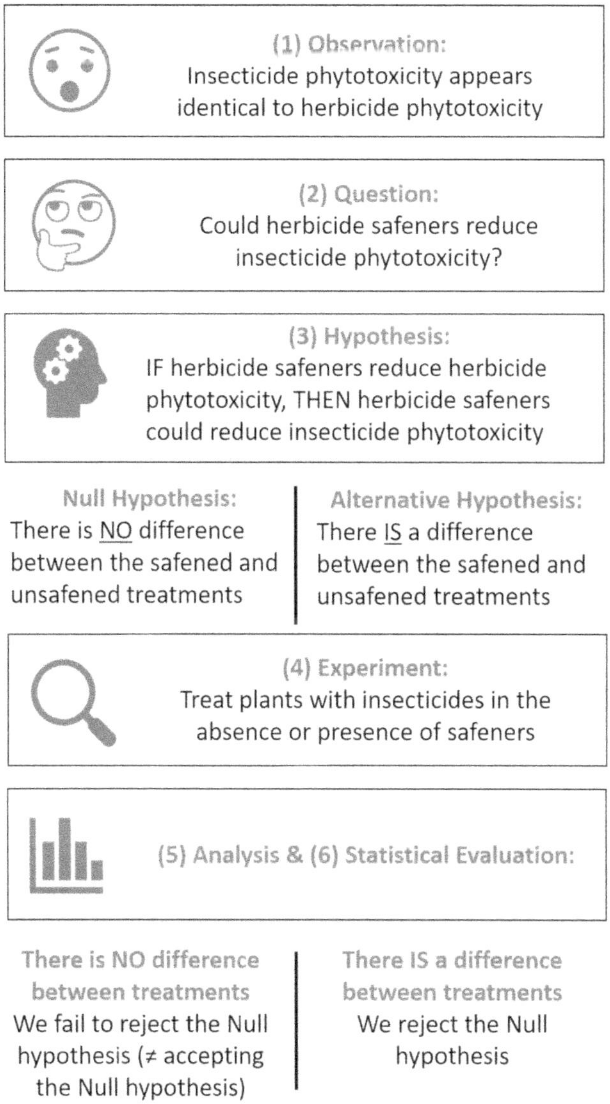

Figure 2.2: The Scientific Method - herbicide safening.

---

1. R.A. Fisher introduced the null hypothesis in "The design of experiments" (1935), noting that the normal hypothesis is never proved or established but is possibly disproved in the course of experimentation".

II

DESIGNING AND IMPLEMENTING EXPERIMENTS (SIGNIFICANCE, POWER AND EFFECT)

3

TYPE I AND TYPE II ERRORS

To experimentally test a hypothesis *"IF herbicide safeners reduce herbicide phytotoxicity, THEN herbicide safeners could reduce insecticide phytotoxicity,"* we might predict (the Null hypothesis) that there is no difference between the safened and unsafened treatments.

Accordingly, we could perform an experiment in which we treat plants with insecticides in the absence or presence of safeners and observe for differences in phytotoxicity between treatments, for which the Null hypothesis is: There is no difference between treatments.

During experimentation, researchers will be exposed to two types of error related to hypotheses, termed Type-I and Type-II errors (originally[1] termed "the first source of error" and "the second source of error":

A *Type-I error* occurs when the Null hypothesis is rejected, even though it is correct. This is the "flash in the pan" type error, where you believe you have discovered something extraordinary, even though it is not (a "false positive").

A *Type-II error* occurs when the Null hypothesis is not rejected, despite it being false. This is the "one that got away" type error, where you miss out on something that really is extraordinary (a "false negative).

Let us consider our hypothesis for an experiment to determine whether there is a difference in phytotoxicity between two treatments. We can formulate this as:

Null hypothesis: There is no difference between treatments.

Alternate hypothesis: There is a difference between treatments.

In this example, a Type-I error would lead us to reject the Null hypothesis, claiming that there IS a difference between treatments (false positive) when there is none.

A Type-II error would lead us to *not* reject the Null hypothesis, erroneously concluding that there is NOT a difference in phytotoxicity between the treatments.

In an early-stage herbicide discovery project focusing on efficacy, a Type-II error (false negative) can significantly impact progress and potentially allow a competitor to commercialize a valuable compound.

For example, imagine a research team screening various compounds to identify potential herbicides, including a promising compound with strong efficacy against a specific weed species. However, due to a Type-II error, they mistakenly conclude that the compound lacks sufficient efficacy and discard it prematurely.

Meanwhile, a competitor's research team obtains access to the compound and recognizes its potential. They conduct further studies, optimize its formulation, and successfully demonstrate its effectiveness. Consequently, the competitor proceeds to commercialize the herbicide, gaining a competitive advantage in the market.

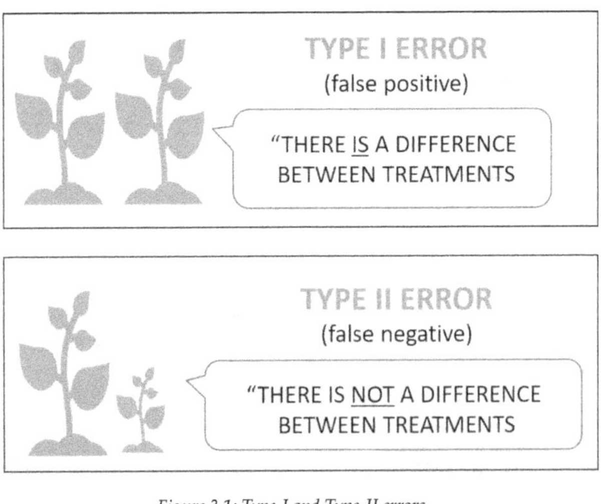

Figure 3.1: Type-I and Type-II errors.

As a result of the Type-II error, the original research team missed the opportunity to recognize the compound's true efficacy and failed to capitalize on its potential.

This can have detrimental consequences for the initial project, as the competitor now has the advantage of a commercially viable herbicide that effectively controls the targeted weed species. The original team may experience setbacks in their research, lose out on potential revenue and market share, and face challenges in catching up with the competitor's success.

In a late-stage herbicide development project focused on efficacy, a Type-I error (false positive) could have significant implications. For instance, imagine a herbicide that has undergone extensive testing and regulatory approvals, with the conclusion that it effectively controls a wide range of weed species.

Due to a Type-I error, the development team mistakenly concludes that the herbicide is highly efficacious against all targeted weeds. As a result, the herbicide moves towards commercialization, and farmers widely adopt it based on the false belief in its effectiveness.

However, if later field trials or real-world applications reveal that the herbicide fails to deliver the expected control against certain weed species, it would be a Type-I error.

The consequences could include crop damage, reduced yields, economic losses for farmers, and loss of trust in the product. Thus, a Type-I error in late-stage herbicide efficacy assessment can have detrimental effects on crop management, farmer livelihoods, and the success of the product.

As we delve into the following chapters, it will become evident that understanding the effects of Type-I and Type-II errors on early-stage discovery and late-stage development projects influences our choices in trial design and the evaluation of trial data.

Recognizing the potential consequences of these errors allows us to make informed decisions that optimize the efficiency and efficacy of our research endeavors.

In the realm of early-stage discovery projects, where the focus lies on identifying novel solutions or opportunities, the risk of a Type-II error (false negative) looms large. By understanding the impact of this error, we become aware of the possibility of missing out on valuable discoveries or dismissing potential breakthroughs prematurely.

Consequently, we can design trials and develop data evaluation methods that are sensitive enough to detect even subtle signals of efficacy or significance, reducing the likelihood of Type-II errors and ensuring that promising leads are thoroughly explored.

Conversely, in late-stage development projects characterized by higher costs and greater emphasis on product viability, a Type-I error (false positive) poses a considerable threat.

The recognition of this error's impact empowers us to approach trial design and data evaluation with a critical eye, striving to minimize the risk of false positives.

We employ more stringent statistical analyses to ensure that any positive findings are truly indicative of efficacy or success. By doing so, we mitigate the potential consequences of Type-I errors, such as wasted resources, misleading conclusions, and misguided decisions regarding the future of the project.

By proactively addressing the specific risks associated with each error, we optimize our chances of identifying valuable discoveries in the early stages and ensuring the reliability of our findings in the later stages.

This understanding ultimately enhances the effectiveness and impact of our research efforts in various domains and facilitates better-informed decisions throughout the project lifecycle.

Assumptions for hypothesis testing	
Type I error (false positive)	Significance (α) 0.05 (5%)
Type II error (false negative)	Power (β) 0.8 (80%)

Although Type I and Type II errors can never be avoided entirely, we can reduce their likelihood. The probability of Type I error is controlled by the significance level (stringency) of the test, denoted by alpha (α), which is typically set at 0.05 (5%).

The probability of Type II error is controlled by the power (sensitivity) of the test, or beta (β), which is typically set at 0.20 (20%) or 80% power.

Two parameters determining appropriate sample sizes will be addressed in the next chapter on *Statistical significance* (for Type-I errors) and *Power* (for Type-II errors).

---

1. The concept of Type-I and Type-II error was originally referred to in 1928 as "the first source of error" and "the second source of error" in papers by J. Neyman and E.S. Pearson (son of Karl Pearson).

4

STATISTICAL SIGNIFICANCE, POWER AND EFFECT SIZE.

In statistical analysis and hypothesis testing, the concepts of significance, variability and effect size play crucial roles in determining the power or sensitivity of a test.

Power refers to the ability of a statistical test to detect a true difference or effect if it exists in the population being studied.

The interplay between effect size, variability, significance, and power determines the sensitivity of a statistical test to detect differences between treatments.

In Figure 4.1, the relationship between effect size, variability, and the ability to differentiate treatments is depicted. When the effect size is suitably large, even if there is substantial variability within the treatments, the treatments can be easily differentiated, making it less challenging to detect a true difference or effect if it exists.

However, as the effect size decreases (a smaller difference between treatments), data overlap increases, and it becomes more challenging to differentiate between the treatments. The statistical test needs to be more sensitive to detect the smaller effect size accurately.

The sensitivity or power of the trial becomes crucial in such cases. A test with higher power is more likely to identify the difference between treatments, even when the effect size is small and the variability is large.

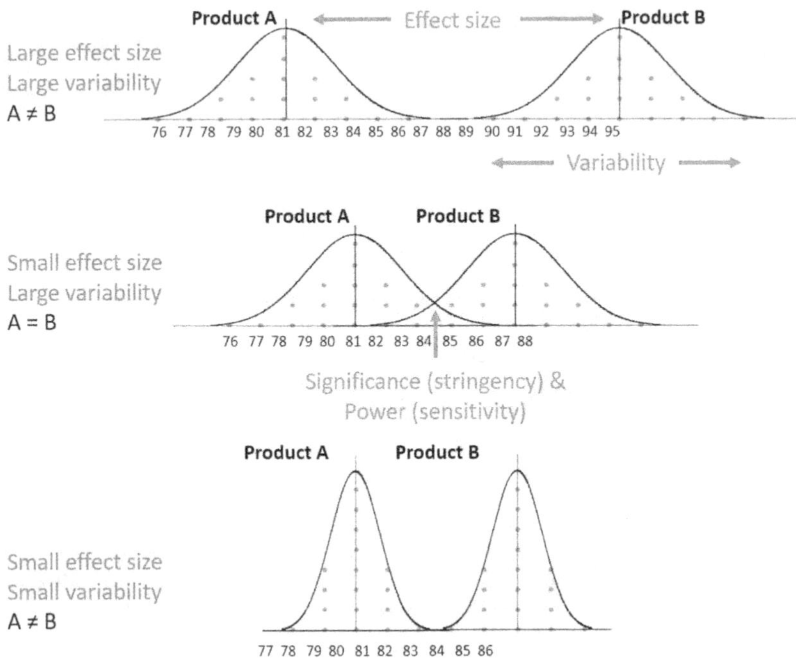

Figure 4.1: The relationship between effect size, variability, and the ability to differentiate treatments.

Significance (stringency)

The significance level of an experiment, called α (alpha) is the probability of rejecting our Null-hypothesis even though it is true, thus concluding that there is a difference between treatments, even though there is none.

The **significance level (α)**, is the probability of rejecting the null hypothesis when it is actually true. For a significance level of 0.05 (or 5%), there is a 5% chance of mistakenly rejecting it and concluding that there is a significant effect or difference (false positive).A small difference in efficacy between two treatments may be statistically significant, but this difference may be too small to make it commercially meaningful!

A lower significance level means that the test is more stringent and less likely to reject the null hypothesis, leading to lower power.

Conversely, a higher significance level means that the test is less stringent and more likely to reject the null hypothesis, leading to higher power.

This is of importance for biological trials where we as researchers are biased toward the conclusion that our experiments support the hoped-for effect - the alternative hypothesis.

Let us consider our proposed experiment in which plants sprayed with a phytotoxic insecticide *with* added herbicide safener is tested, relative to plants sprayed with a phytotoxic insecticide *without* added safener.

Our experimental data reveals that for 20 sprayed plants, all 20 show phytotoxicity in the absence of the safener, while only 19 of those sprayed with the safener show phytotoxicity. Clearly, the one symptom-free plant could result from chance, and there would be no reason to reject the Null hypothesis.

If, however, 19 of the safener treated plants showed no sign of phytotoxicity, common sense would lead us to conclude that this was extremely unlikely to happen due to chance and we would reject the Null hypothesis, concluding that the treatment had an effect.

But what if only nine plants showed no sign of phytotoxicity - here, more than common sense is required to determine the likelihood of this occurring due to chance, and we will need to calculate the probability of the Null hypothesis being rejected - and that the safener treatment *does* have a significant effect.

A biological rule of thumb is that a 5% significance level is considered appropriate, meaning that we can be 95% confident that there is, in fact, a difference between treatments.

Stated in statistical terms, the typical significance level (alpha) of 0.05 (5%) corresponds to a 95% confidence level - we are 95% confident that we will not make a Type-I error (identifying false positives) and reject the Null hypothesis despite its being true.

In other words, only 5% of our experiments will give us a false positive. This level is termed the "*critical value*". If the test statistic falls below the critical value (in this case α=0.05), we accept the null hypothesis (H0), if it falls above the critical value, we reject the null hypothesis.

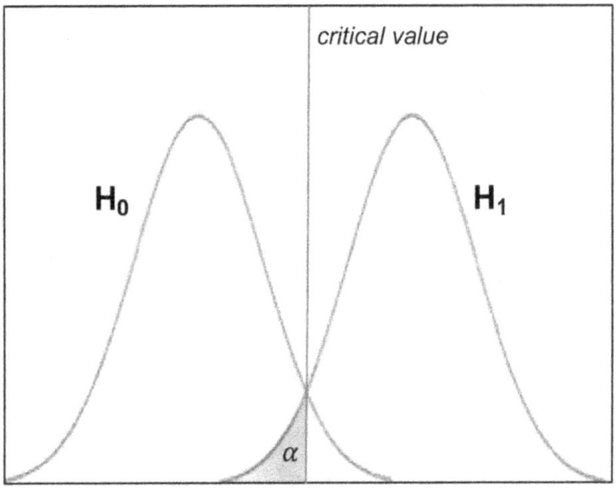

Figure 4.2: Significance level as determined by the critical value. If the test statistic falls below the critical value, we accept the null hypothesis (H0). α (alpha) represents the probability of making a Type-I error: rejecting our Null-hypothesis even though it is true (false positive).

The 5% significance level is a threshold arbitrarily decided on by biologists and is the threshold most commonly used in the Biological Sciences. Depending on the trial circumstances, *i.e.* what the data is to be used for, a 1% or 10% significance level may be appropriate.

Significance Rule of Thumb II: if the cost of a false positive is high (leading to a strategic decision to initiate a costly development process) thresholds for critical trials may be more stringent (1%). Conversely, if the cost of a false negative is high (leading to you missing out on a valuable discovery), the threshold for initial screening experiments may be raised to e.g. 10%.

If the cost of a false positive is high (for example leading to a strategic decision to initiate a costly development process) thresholds for critical trials may be made more stringent (1%).

Conversely, if the cost of a false negative is high (leading to you missing out on a valuable discovery), the threshold for initial screening experiments may be raised to, *e.g.* 10%.

If the probability (p-value, or "p") that the test statistic calculated from the observed data occurred by chance is less than α = (5%), we may reject the Null hypothesis. This is commonly phrased as "a small p-value indicates that the Null hypothesis is not a good explanation for the data - we can be 95% confident that the alternative hypothesis is true".

This opens for one of the most hotly debated topics in biological statistics - the misuse of the p-value. Statisticians argue that it is incorrect to transpose the observation of a "5% chance of getting the observed results if the Null hypothesis is correct" into an observation of a "95% probability that the Null hypothesis is false" or "95% certainty that the observed difference is real and could not have arisen by chance" or "the difference is statistically significant".

It is beyond the scope of this book to expand on this debate[1] but suffice to say the above transposition is commonly used in the biological sciences, and the following paragraphs should be read with the above in mind.

It is common practice in most of the sciences to provide a p-value resulting from a statistical test and on this basis to conclude that the results are significant or not significant.

As seen above, the 5% significance level is a threshold arbitrarily decided on by biologists.

p-value	Graphical representation	Level of significance
<0.001	***	Very highly significant
<0.01	**	Highly significant
<0.05	*	Significant
<0.1		Approaching significance
0.1 or more		Not significant - more than 10% probability of chance

Figure 4.3: p-values and associated levels of "statistical significance".

Almost immediately, it appears, biologists (probably those that had p<0.06% datasets!) began to argue that fixed significance levels were too restrictive for biological data sets and devised a system of graduated levels of "the difference is statistically significant".

This tendency to correlate p-values with significance should be tempered with the consideration of whether the observed differences are substantively important.

For example, a small difference in efficacy between two treatments maybe statistically significant according to the above (hypothesis p-value) definition, but this difference may be too small to make it commercially meaningful (biologically significant).

For this reason, statisticians typically recommend estimating the effect size and evaluating confidence intervals for the differences between means, to emphasize the importance of the *magnitude* of the effect rather than merely statistically significant/non-significant hypothesis testing. These parameters will be discussed in the following.

Effect Size

Statisticians define *effect size* as the minimum acceptable deviation from the Null hypothesis and is a quantitative measure of the magnitude of the difference between the two groups, H0 and H1:

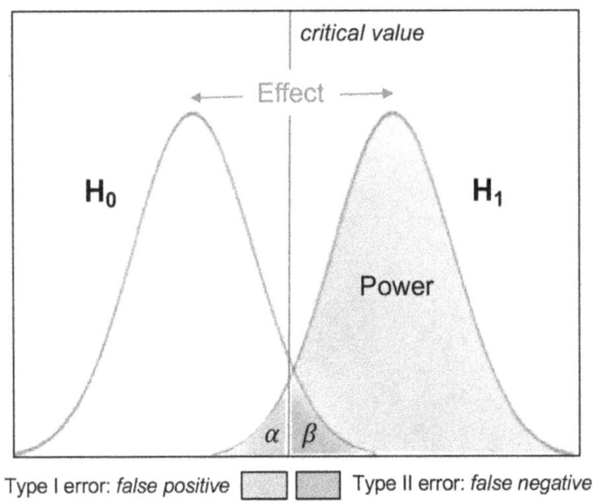

Figure 4.4: Effect size is a measure of the magnitude of the difference between H0 and H1 and is often defined as the minimum improvement in efficacy needed to justify the costs of developing a new product.

For crop protection researchers, effect size (or treatment effect) is often defined as the minimum improvement in efficacy needed to justify the costs of developing a new pesticide formulation.

A biological rule of thumb is that an effect size (treatment effect or improvement in efficacy) of 20% is considered economically viable.

As we will see shortly, effect size can have a significant influence on the number of replicates needed to perform the experiment. The trade-off is cost and space; it is not always feasible to include the requisite number of replicates in, for example, a greenhouse experiment.

Power or Sensitivity of a Test

The *power or sensitivity of a test* is used to determine the appropriate sample size for a test or experiment.

Before planning an experiment, researchers must consider which level of statistical power is sufficient to ensure that the test is sensitive enough to identify the difference between treatments.

Stated in statistical terms, *Power is the probability of correctly rejecting the Null hypothesis when it is false* and identifying a significant effect when such an effect exists.

The probability of Type II error is controlled by the power (sensitivity) of the test, denoted by beta (β), which is typically set at 0.10 or 0.20 (respectively a 90% (or 80%) chance of identifying a significant effect when such an effect exists.

A biological rule of thumb is that a Power of 80% is considered appropriate, meaning that there is only a 20% chance of erroneously concluding that there was no difference in efficacy between the treatments.

The trade-off is that the principal way to increase statistical power is to increase the number of replicates or the number of treatments - depending on the variability of biological material. This can require so many replicates that the trial becomes economically and practically unfeasible.

Researchers need to calculate the lowest sample sizes which will permit a statistically viable experiment.

To be able to do this, the statistical power of the test must be defined, and the *variability* or *variance* within the experimental setup must be determined.

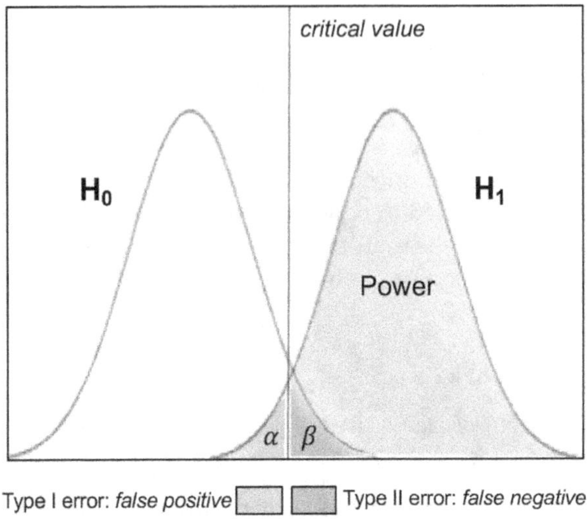

Figure 4.5: Power (or sensitivity of a test) is the probability of correctly rejecting the Null hypothesis when it is false and identifying a significant effect when such an effect exists. α (alpha) represents the probability of making a Type-I error: rejecting our Null-hypothesis even though it is true (false positive). β (beta) represents the probability of making a Type-II error: accepting our Null-hypothesis even though it is false (false negative).

Factors determining whether the test is sensitive enough to identify the difference between treatments, and to control the probability of Type I and Type II errors are summarized here:

Significance level: The significance level (alpha) is the probability of rejecting the null hypothesis when it is actually true. A lower significance level means that the test is more stringent and less likely to reject the null hypothesis, leading to lower power. Conversely, **a higher significance level means that the test is less stringent and more likely to reject the null hypothesis, leading to higher power.**

Variability: The variability in the data is the degree to which the observations or measurements differ from each other. **A higher variability generally leads to lower power because it makes it more difficult to detect a true effect or difference.** If the variability is large, the test may not be able to distinguish between the true effect and random variation, leading to lower power.

Effect size: The effect size is the magnitude of the difference or relationship between the groups or variables being compared. **A larger effect size generally leads to higher power because it makes the difference between the groups or variables easier to detect.** If the effect size is small, the test may not be able to detect the difference between the groups or variables, leading to lower power.

Sample size: The sample size is the number of observations or participants in a study. **A larger sample size generally leads to higher power** because it reduces the sampling error and increases the precision of the estimate. The larger the sample size, the more likely the test is to detect a true effect or difference, leading to higher power.

---

1. Although the p-value was introduced by Karl Pearson, it was extensively applied and disseminated in R.A. Fishers "Statistical methods for research workers" (1925). Here, he advocated the 5% significance level, (identifying that 1.96 standard deviations around the mean is the approximate value of the 97.5 percentile point of the normal distribution) which in turn led to the rule of thumb of two standard deviations for statistical significance in a normal distribution. One cannot help noticing the irony in light of the ongoing (often heated) debate regarding the use of p-values rather than fixed significance levels!

5

VARIANCE

An experiment comprises an independent as well as a dependent variable. The independent variable (in our example, the herbicide safener) can be changed or controlled (included or eliminated), and the effect it has on the dependent variable (insecticide phytotoxicity) can be recorded and analyzed.

A third group of variables are termed controlled variables and are those variables which the researcher controls (holds constant) and quantifies (records) during an experiment.

Temperature is a typical controlled variable - as temperature could affect the degree of phytotoxicity, it is important that it is held constant during the experiment. Other controlled variables relevant for biological trials include light, humidity and the variability within laboratory equipment.

Variance

Variance is the final factor which determines the required sample size (number of treatments or number of replicates) when designing an experiment. For the estimation of variance in a specific trial, the experimental standard deviation will need to be determined.

Variance is a measure of how far a data set is spread out or differs from the mean value. Variance is defined as the average of the squares of the differences between the observed values and the calculated average.

A biological data set will typically have a high variance, and experimental variance may arise from a number of sources including variance within the biological material being tested, researcher variance, equipment variance, and climatic variance.

Standard Deviation

Standard deviation (SD) is calculated as the square root of the variance, and approximates the average difference, or "spread" of the data set from the mean.

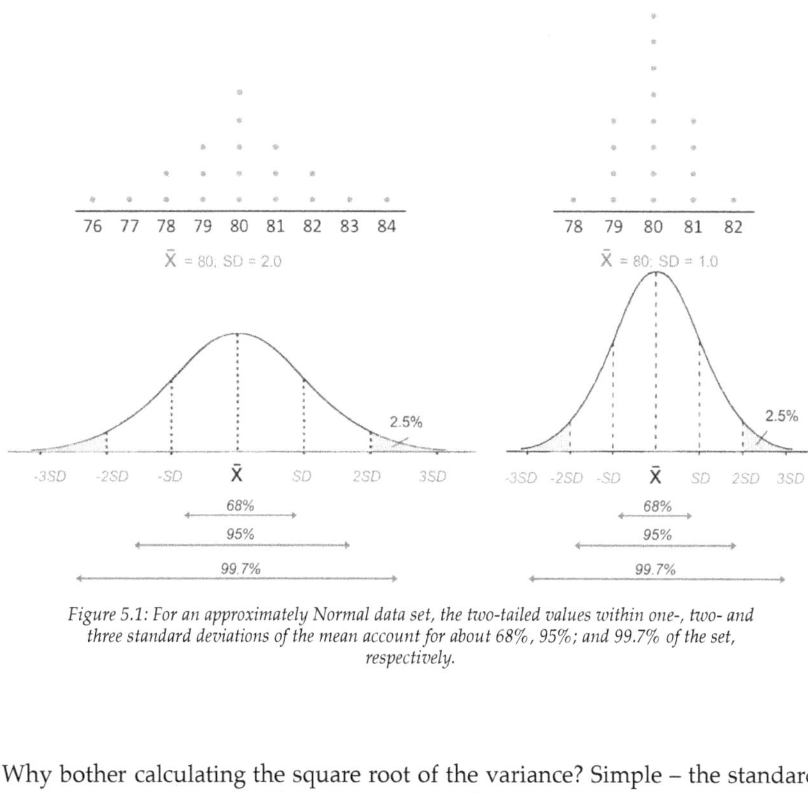

Figure 5.1: For an approximately Normal data set, the two-tailed values within one-, two- and three standard deviations of the mean account for about 68%, 95%; and 99.7% of the set, respectively.

Why bother calculating the square root of the variance? Simple – the standard deviation is expressed in the same units as the mean ($\bar{x}$), whereas variance is expressed in squared units. Standard deviation is thus a more intuitive method of indicating the spread of our data around the mean.

From figure 5.1, we can see that about 68%, 95% and 99.7% of the values in a Normal distribution lie within one, two and three standard deviations of the mean, respectively. For practical purposes, scientists assume that their data is derived from an approximately Normal data set.

One-tailed vs. two-tailed tests

In Figure 5.1, the standard deviation intervals are based on a two-tailed (or two-sided) test, which distributes the possibility of an effect in both directions: positive and negative.

Stated in other terms: a two-tailed test divides your significance level, and each direction is only half as "strong" as a one-tailed test, in which the significance level is applied in a single direction.

Without going into more detail, it is sufficient to know that for biological testing, two-sided tests should be used, unless there is reason (and you have the statistical insight) to do otherwise.

Two-sided Rule of Thumb: for the biological sciences, two-sided statistics should be used, unless there is reason (and you have the statistical insight) to do otherwise.

Normal distribution

Normal data distributions are symmetrical and follow a bell-shaped density curve (see Figure 5.1), with data distribution denser in the center (around the mean) and less dense in the tails.

Normality Rule of Thumb: for biological data, a Normal distribution may be assumed as a valid working approximation.

For biological data, the assumption of Normality is generally a valid approximation, as the Normal distribution of biological data is surprisingly ubiquitous. If the data is clearly not Normally distributed, it may need to be transformed - this will be discussed in a later chapter.

Estimating Standard Deviation

Standard deviation may be estimated from pilot trials or similar, prior experiments. Often as a researcher, you will be handed a set of data from a similar trial and expected to give an on the spot evaluation of its variance.

A useful trick to have up your sleeve is the ability to give a rough estimate of the standard deviation of a data set using the Range Rule of Thumb:

Range Rule of Thumb: SD ≈ Range/4 where Range = (maximum value) - (minimum value).

The Range Rule becomes more accurate when written as SD ≈ Range / SQRT(n). So, to be able to give a quick estimate of the standard deviation by glancing at a table of data, we may use the Range Rule for estimating Standard Deviation (SD):

SD ≈ Range/SQRT(n)	
SD estimate	**Dataset**
SD ≈ Range/2	<7
SD ≈ Range/3	8 - 12
SD ≈ Range/4	12 - 20

For rapid estimates of Standard Deviation in efficacy (0-100%) datasets, the following graphical representation provides a useful reference:

		Data set size (n)		
		<7	**8-12**	**12-20**
Range	**60**	30	20	15
	55	28	18	14
	50	25	17	13
	45	23	15	11
	40	20	13	10
	35	18	12	9
	30	15	10	8
	25	13	8	6
	20	10	7	5
	15	8	5	4
	10	5	3	3
	5	3	2	1

Figure 5.2: Standard Deviation estimates for selected ranges and data set sizes.

Let us consider our proposed experiment in which plants sprayed with a phytotoxic insecticide with added herbicide safener is to be tested, relative to plants sprayed with a phytotoxic insecticide without added safener.

In a previous pilot trial using the phytotoxic insecticide against the same plant species we intend to use, the results of 19 replicates were distributed as shown in Figure 5.1 (left).

The range is thus 84 - 76 = 8 and applying the Range rule of Thumb allows us to quickly estimate SD as 8/4 = 2. In this example, a calculation of SD gives the same result.

We will examine the usefulness of the standard deviation, Coefficient of Variation and effect size in more detail in the later section: *Descriptive Statistics - Critically evaluating experimental data.*

6

SAMPLE SIZE & REPLICATION

Replication is the repeated application of treatments to multiple independently assigned experimental units. The number of independently assigned experimental units that receive the same treatment is the sample size.

To obtain a rough estimate of the number of replicates required to generate a data set with a specific effect size, significance level and power, the estimated coefficient of variation for the experiment first needs to be determined:

Coefficient of Variation

Experimental variability may be expressed as the percent data spread relative to the mean and is termed the Coefficient of Variation or Coefficient of Variance (CoV), also known as the Relative Standard Deviation (RSD).

Calculation of the CoV should only be performed on data measured on the ratio scale (*e.g.*, mass, length, duration, disease severity, etc.), and on Normally distributed data (see previous chapter).

Also, it should be noted that when the mean value is small (approaching zero) the CoV may become extremely large and will be sensitive to small changes in the mean. Accordingly, CoV for *e.g.* crop protection trials are typically calculated for the ED50 (the dose giving a 50% effect).

The Coefficient of Variation may be calculated as:

$$CoV = (SD/MEAN)*100$$

The Coefficient of Variation describes the size of the standard deviation or data spread relative to the mean (or average). Standard Deviation and Mean may be estimated from pilot trials or similar, prior experiments.

Accordingly, in our proposed experiment in which plants sprayed with a phytotoxic insecticide with and without added herbicide safener is to be tested, the results of an initial pilot trial indicate that the data range was thus 84% efficacy (highest value) - 76% efficacy (lowest value) = 8 (see previous chapter).

Applying the Range rule of Thumb allows us to estimate SD as 8/4 = 2. From this we may estimate the Coefficient of Variation as CoV = (2/80)*100 = 2.5%

A biological rule of thumb is that a coefficient of variation of 20% or less is considered acceptable.

As variance increases for an experiment, it becomes more difficult to detect a significant difference, and a larger sample size will be required. In the absence of pilot trials or similar prior experiments, assumptions of CoV may be used together with Power and Effect Size (see earlier chapters) for a rough estimate of required sample size when designing experiments.

Biological assumptions for estimating number of replicates

- a variability (CoV) of 20% or less is considered acceptable.

- a α=5% significance level is considered appropriate, meaning that only 5% of our experiments will give us a false positive. If the cost of a false positive is high, thresholds for critical trials may be made more stringent (1%). Conversely, if the cost of a false negative is high, the threshold for initial screening experiments may be raised to, e.g. 10%.

- an effect size (treatment effect or efficacy improvement) of 20% is considered economically viable (RRDI).

- a Power of 80% is considered appropriately sensitive, meaning that there is a 80% chance of identifying a significant effect when such an effect exists - the test is sensitive enough to identify the difference between treatments.

Assumptions for Design of Experiments (DOE)	
Significance (stringency)	5% (10)
Variability (CoV)	≤20%
Effect (difference)	20%
Power (sensitivity)	80% (90)

Figure 6.1: Assumptions for Design of Experiments (DOE).

With an indication of the relative data spread or variability, an estimate for the number of replicates required to achieve typical levels of significance, power and effect size may be obtained from the following tables, for a single sample (*e.g.* is it true that the treatment effect is greater than 0, the expected effect of an untreated control, UTC).

For comparing two treatments (*e.g.* treated and untreated control), the following table may be used to conveniently estimate the number of replicates needed based on CoV and Effect size, assuming 80% power and 5% significance:

Consider our proposed experiment in which plants sprayed with a phytotoxic insecticide with and without added herbicide safener is to be tested.

☑ Development

α: 5%

% variability in Observations (CoV)	Effect Size 10	15	20	25	30
5	4	3	3	3	3
10	14	6	3	3	3
15	32	14	7	4	3
20	58	24	13	8	5
25	90	38	20	12	8
30	130	55	29	17	11
35	177	74	39	24	15
40	231	97	51	31	20

☑ Discovery

α: 10%

% variability in Observations (CoV)	Effect Size 10	15	20	25	30
5	3	3	3	3	3
10	7	3	3	3	3
15	16	7	4	3	3
20	29	12	6	4	3
25	45	19	10	6	4
30	65	27	14	9	6
35	88	37	20	12	8
40	115	48	26	15	10

Figure 6.2: Sample sizes for a range of COV and effect sizes for comparing two treatments, two-sided and assuming 80% power and 5% significance (for Development trials) and 10% (for Discovery trials. Based on "Statistical Rules of Thumb" by Gerald van Belle (2008).

In the absence of any pilot trial data, or data from similar trials we may start with some underlying assumptions: a treatment effect of 20% (the safener-treated plants will need to show at least 20% less phytotoxicity to make further development worthwhile, and an assumption of 20% variability (CoV) is a realistic estimate of variability. We can now roughly estimate that each of the treatments will require 13 replicates (Figure 6.2).

For trials with a greater number of treatments or "experimental units", the number of replicates required may be determined from the **minimum accepted degrees of freedom for a biological trial**. This will be covered in the next chapter.

7

EXPERIMENTAL DEGREES OF FREEDOM

For trials with a greater number of treatments or "experimental units", the number of replicates required may be determined from the **minimum accepted degrees of freedom for a biological trial**.

Degrees of freedom refers to the number of independent pieces of information used to calculate the statistic. The degrees of freedom are calculated by subtracting one from the number of samples in your data set.

Simply put - for a one-sample statistical test, *i.e.* two treatments (one herbicide sample plus an untreated control), one degree of freedom is spent estimating the mean of the sample, while the remaining n-1 degrees of freedom are available to estimate variability.

When comparing two treatments, a larger number of replicates per treatment is required to achieve sufficient statistical power because the variability within each treatment needs to be estimated accurately.

This ensures that any observed differences between the two treatments are not due to random chance or natural variability. By increasing the number of replicates per treatment, the estimates of variability become more reliable, increasing the likelihood of detecting a true difference if it exists.

On the other hand, when comparing multiple treatments, the variability within each treatment is still important, but the focus shifts more towards capturing the overall variation between the treatments.

Since the variability between treatments tends to be larger than the variability within treatments, fewer replicates per treatment are needed to detect meaningful differences between the treatments.

This is because the increased variability between treatments provides more information for statistical analysis and increases the chances of detecting true differences.

The minimum accepted number of degrees of freedom required for a trial design to be considered adequate is 12 (*EPPO Guidelines, Design and analysis of efficacy evaluation trials - EPPO PP 1/152(4)*).

A biological rule of thumb is that the minimum accepted number of degrees of freedom required for a trial design is ≥12

However, this should be increased if there is low precision in the measurements taken, and for biological trials, a minimum residual degree of freedom of 15 is considered appropriate for a useful statistical analysis.

The EPPO Guidelines note that the actual number of degrees of freedom needed for a given trial may depend on various factors, including the expected variability of the response variable, the magnitude of treatment effects, and the desired level of statistical power.

Therefore, researchers are advised to carefully consider these factors when planning their trials and adjust the sample size accordingly to achieve an appropriate number of degrees of freedom.

The EPPO Guidelines do not explicitly provide specific values for the variability of the response variable, the size of treatment effects, or the desired level of statistical power that underlie the recommendation for a minimum of 10 degrees of freedom per treatment group.

However, if we extrapolate the data provided in the guidelines, it appears they are based on the *"Assumptions for Design of Experiments (DOE)"* presented in the previous chapter.

One-sample degrees of freedom

For an efficacy data set for one herbicide with ten replicates, the degrees of freedom will be:

10df (1 sample x 10 replicates) - 1 sample df = 9df

Here the degrees of freedom may be increased by increasing the replication. If this is not possible, degrees of freedom may be increased by increasing the number of treatments or the number of sites (replication of the experiment).

• • •

Two-sample degrees of freedom

For a two-sample t test, *i.e.* three treatments (two herbicide samples plus an untreated control) one degree of freedom is spent estimating each mean, while the remaining n-2 degrees of freedom are available to estimate variability.

If you are comparing the efficacy of two herbicides at comparable doses (*i.e.* 3 treatments: two herbicides plus the untreated control), with 10 replicates for each treatment, the degrees of freedom will be:

20df (2 samples x 10 replicates) - 2 sample df = 18df

Degrees of Freedom for a single, completely randomized site

Degrees of Freedom for a single, completely randomized site (*e.g.* a typical greenhouse trial) for selected combinations of Replicates and Treatments may conveniently be obtained from the following table (Figure 7.1).

From this table we can see that for a trial with 4 treatments (3 herbicide samples plus an untreated control) and 3 replicates, there are only 6 residual df. We will thus need at least 5 replicates to obtain df=12, while 6 replicates to obtain df=15 would be preferred.

		Replicates							
		3	**4**	**5**	**6**	**7**	**8**	**9**	**10**
Treatments	**2**	2	3	4	5	6	7	8	9
	3	4	6	8	10	12	14	16	18
	4	6	9	12	15	18	21	24	27
	5	8	12	16	20	24	28	32	36
	6	10	15	20	25	30	35	40	45
	7	12	18	24	30	36	42	48	54
	8	14	21	28	35	42	49	56	63
	9	16	24	32	40	48	56	64	72
	10	18	27	36	45	54	63	72	81

Figure 7.1: Degrees of freedom for replicates and treatment numbers, single site (5% significance).

Alternatively, we could repeat this trial at an additional site - doubling the df to 12, or (as is common for crop protection field trials) at three sites, tripling the degrees of freedom to 18.

From this table we can see that for a trial with 4 treatments (3 herbicide samples plus an untreated control) and 3 replicates, there are only 6 residual df. We will thus need at least 5 replicates to obtain df=12, while 6 replicates to obtain df=15 would be preferred.

Now that we have determined how large our trial needs to be (sample size; replicates), this would be an excellent time to head out into the greenhouse and consider how our trial is to be set up! Once we have identified the number of replicates to be used in our experiment, we may now consider how to assign and distribute (randomize) our experimental units.

8

RANDOMIZATION AND EXPERIMENTAL DESIGN

One of British statistician and geneticist R.A. Fisher's (1892-1962) most important contributions to statistical methods is the introduction of randomization to experimentation, providing solutions to issues of variance arising from factors such as the heterogeneity of soils and the variability of biological material.

In biological trials, randomization is the process of randomly allocating experimental units (*e.g.* pots comprising treatments, references, untreated controls, as well as application timings and -doses) across the treatment groups.

Randomization provides a basis for the statistical methods used in generating and analyzing data by reducing bias, such as selection bias (our tendency to see patterns when evaluating experiments).

Controls

Scientific controls are part of the scientific method. Untreated controls are always included when designing experiments to confirm that observed effects are in fact due to the treatment. To control for confounding variables, such as the physical effects of spraying plants, untreated controls in crop protection trials are typically sprayed with water, or with a solution of formulation compounds from which the active ingredient has been eliminated.

For crop protection trials, the principal randomized designs in common use are the Completely Randomized Design (CRD) and the Randomized Complete Block Design (RCBD).

The Completely Randomized Design (CRD)

In a Completely Randomized Design (CRD) treatments are distributed at random. Each experimental unit (such as a plot of land or a group of plants) is randomly assigned to one of the treatment groups.

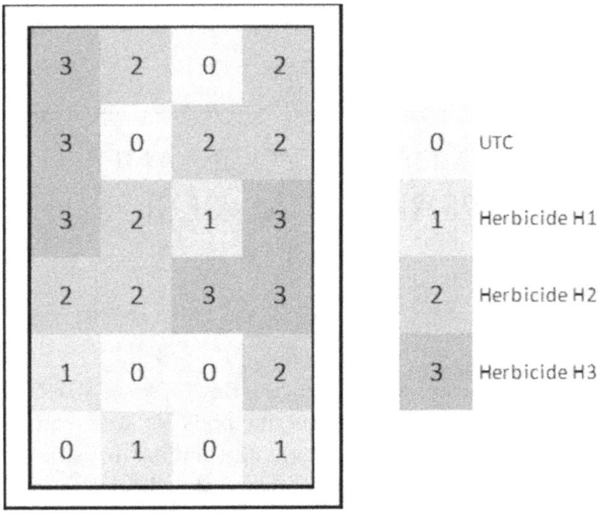

Figure 8.1: Completely Randomized Design (CRD) for four treatments (one untreated control, UTC and three herbicide samples).

This means that each experimental unit has an equal chance of receiving any particular treatment. CRD is relatively straightforward to implement, especially when the number of treatments is small and the experimental units are homogeneous.

Carried out on a greenhouse table, the pots could be randomly placed as shown in Figure 8.1. For a trial with four treatments (one untreated control, UTC and three herbicide samples, H1, H2 and H3), we can see from the previous section that we need six replicates to attain 15df.

The Randomized Complete Block Design (RCBD)

Completely Randomized Design (CRD) assumes that conditions across the greenhouse table (or field) are equivalent.

In reality this is rarely the case, and we can expect variation in temperature, light and humidity depending on the experimental units (pots) proximity to windows, insulated walls, etc.

These differences may be reduced by optimal greenhouse design, but their presence and influence on plant growth and response may also be addressed statistically by restricting the randomization process to form blocks.

The goal of blocking is to reduce the variability caused by factors that are not of primary interest in the study.

For example, if different sections of a field have different soil fertility levels, blocking can help ensure that each treatment is represented equally across those sections.

By incorporating blocking into the experimental design, RCBD provides more precise estimates of treatment effects and enhances the ability to detect smaller differences between treatments.

These blocks, containing all the treatments, treatment levels or combinations (called complete blocks) are typically set up at right angles to variance gradients.

In Figure 8.2, we take a temperature gradient (due to proximity to uninsulated glass panels and insulated interior walls) into account by placing blocks at right angles to the gradient.

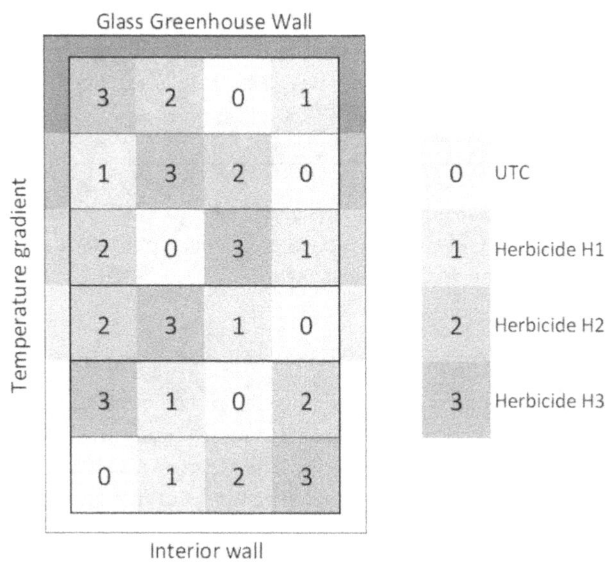

Figure 8.2: Randomized Complete Block Design (CRD) for four treatments (one untreated control, UTC and three herbicide samples).

Each treatment appears in each (complete) block, and are assigned at random within the block, and for 6 replicates we have six blocks.

It is common practice in crop protection trials to ensure (for demonstration purposes) that a single block retains the treatment order as described in the experimental protocol.

RCBD enhances the ability to detect smaller differences between treatments, especially when there is a need to control specific factors. The Randomized Complete Block Design (RCBD) is more precise than the Completely Randomized Design (CRD) because it accounts for variability among experimental units by grouping them into blocks.

Each block contains a complete set of treatments, therefore differences among blocks are not due to treatments, and this variability can be estimated as a separate source of variation. The removal of an appreciable amount of this source of variation reduces experimental error and improves the ability of the experiment to detect smaller treatment differences.

Completely Randomized Design (CRD) is a design where each experimental unit (e.g. plot or a group of pot) is randomly assigned to one of the treatment groups, so each unit has an equal chance of receiving a specific treatment.

Randomized Complete Block Design (RCBD) incorporates blocking to reduce variability caused by factors not of primary interest, resulting in more precise treatment effect estimates when there is significant variability or a need to control specific factors.

In addition to Randomized Complete Block Design (RCBD) and Completely Randomized Design (CRD), there are several other options for randomization in experimental crop protection trials:

A **Latin square design** is used when there are two blocking factors. The treatments are arranged in a square such that each treatment occurs once in each row and column. This design is useful when you want to control for two sources of variability (such as temperature gradient and plant variety) that may affect the response variable.

In **split-plot designs**, when some factors are harder to vary than others at the level of experimental units, it is more efficient to assign the difficult-to-randomize factors (such as irrigation) to larger units (whole plots) and then apply the easier-to-randomize factor (pesticide treatment) to smaller units (subplots). Suppose you want to test the effect of irrigation amount and pesticide type on crop yield. Varying the level of irrigation is difficult on a small scale, so it makes more sense to apply irrigation levels to larger areas of land. You could divide each field into two large plots (whole plots) and apply irrigation amounts to each plot randomly. Then, divide each of these large plots into smaller plots (subplots) and apply pesticide randomly within the whole plots.

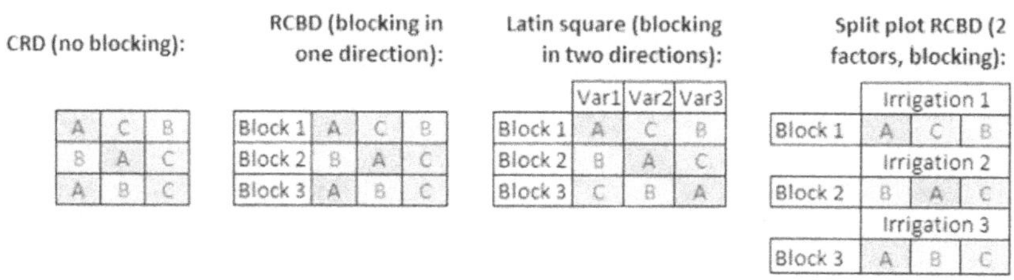

Figure 8.3: Overview of commonly used Randomization options.

These designs offer flexibility in accommodating specific experimental requirements and can enhance the precision and efficiency of crop protection trials by effectively managing various sources of variation.

Randomization offers researchers various approaches to manage sources of variation and increase the validity and precision of their results. By selecting an appropriate randomization method based on the specific characteristics and requirements of the study, researchers can ensure unbiased comparisons and draw robust conclusions about the efficacy of crop protection treatments.

With this information on Experimental Design, and the information on the Scientific Method and Hypotheses as well as Experimental Parameters presented in previous chapters - we are ready to initiate our experiment and move on to the next section: Descriptive Statistics - Critically evaluating experimental data.

III

DESCRIPTIVE STATISTICS - CRITICALLY EVALUATING EXPERIMENTAL DATA

9

DESCRIPTIVE STATISTICS: SD, SE OR 95%CI?

Once we have identified possible outliers (due to biological variance or experimental malfunction) and eliminated them as appropriate (see previous chapter), we are ready to continue the critical evaluation of our experimental data in the form of Descriptive Statistics.

Descriptive Statistics are used to describe the basic features of our experimental data and, together with graphical representations, can provide simple summaries about the Variability, Means, and Significance of our data.

The most common descriptive statistics used in the biological sciences are those of sample Standard Deviation (SD), Standard Error of the Mean (SEM) and 95% Confidence Intervals (95%CI).

Standard Deviation (SD)

Standard Deviation is used to quantify variability in the experimental data and is expressed in the same units as the data.

While calculators and spreadsheets can calculate Standard Deviation at the click of a button, it is occasionally useful to be able to give a rough estimate of the Standard Deviation of a data set using the **Range Rule of Thumb:**

SD ≈ Range/SQRT(n)	
SD estimate	Dataset
SD ≈ Range/2	<7
SD ≈ Range/3	8 - 12
SD ≈ Range/4	12 - 20

Figure 9.1: Range Rule of Thumb to estimate the Standard Deviation of a data set.

This rule of thumb provides a useful estimation for samples sizes greater than n=12. For many biological trials, sample sizes are smaller, and division by 3 (for n = 8-12) or 2 (for n < 7) provides a better estimate (see earlier chapter).

Standard deviation (SD) is calculated as the square root of the variance, and approximates the average difference, or "spread" of the data set from the mean. By squaring the variance the standard deviation is expressed in the same units as the mean. Standard deviation is thus a more intuitive method of indicating the spread of our data around the mean.

Population standard deviation and sample standard deviation are both measures of how spread out a set of data is, but they are calculated differently and are used to describe different groups of data.

The difference between the population and the sample standard deviation is that when calculating the sample standard deviation, we divided by n-1 instead of n to correct the tendency to underestimate the true variability in the population.

Population standard deviation is the measure of how much the data points in an entire population deviate from the population mean. Sample standard deviation, on the other hand, is the measure of how much the data points in a subset, or sample, of the population deviate from the sample mean.

For crop protection trials, we may wish to calculate the standard deviation of the height of ten plants in a treatment. Here we should use the population standard deviation, as we are only interested in the heights of the ten plants in our treatment, and not the plants in any other treatment.

The Standard Deviation cannot be used to determine whether the difference between Means is significant and can only be used to express variability, or to derive the Standard Error of the Mean or its extension, the 95% Confidence Interval.

An example of a practical application of quantifying variability in a trial is given in figure 9.2, where it is apparent that the variability is different for each data point.

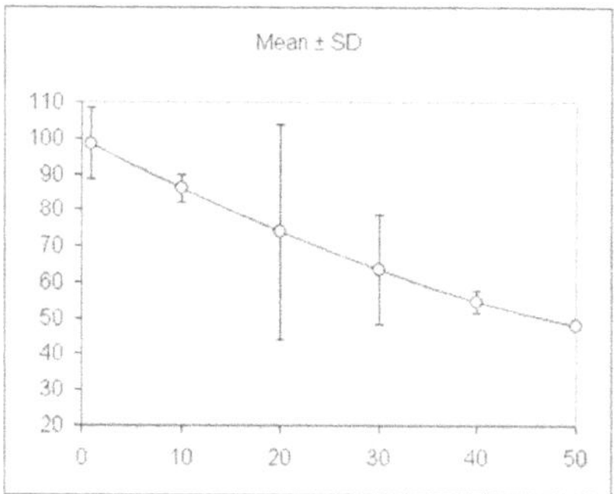

Figure 9.2: Differences in trial variability, expressed as Mean ± SD.

Accordingly, we can deduce that some independent factor (variance within the biological material being tested, researcher variance, equipment variance or climatic variance) contributed to variability and that we should return to our experimental setup to identify and remove this disturbance.

If you do not address this yourself, a reviewer will almost certainly insist that this experiment is repeated, leading to a significant delay in the publication of *e.g.* a scientific article.

Coefficient of Variance (CoV)

We saw in the previous section that as variance increases for an experiment, it becomes more difficult to detect a significant difference, and a larger sample size will be required. We learnt that Experimental variability may be expressed as the percent data spread relative to the mean (the signal:noise ratio) and is termed the Coefficient of Variation or Coefficient of Variance (CoV), also known as the Relative Standard Deviation (RSD).

$$CoV = (SD/MEAN)*100$$

From the perspective of critically evaluating experimental data, we have the:

Variability Rule of Thumb: a coefficient of variation of 20% or less is considered acceptable.

Before initiating a trial, the COV for the specific trial type should be determined, and the trial should only commence if the COV is considered acceptable. On critical evaluation of the trial data, it should be determined whether the data conforms to the agreed COV.

Evaluating Homogeneity of Variance

For two datasets, the **F-test** (next section) can be used to compare the variability of samples, to determine whether they have equal or different variances.

If you have access to a statistical package, you can use **Bartlett's test or Levene's test** to assess the equality of variances across multiple groups or samples.

Bartlett's test performs well with larger sample sizes, and is more sensitive to deviations from normality, while Levene's test is suitable for both small and large sample sizes and is less sensitive and more robust in the presence of non-normality.

Alternatively, you can critically evaluate your data by comparing the variability of samples using the **Homogeneity of Variances Rule of Thumb**: if the largest variance is no more than 4 times the small variance (i.e., if the largest standard deviation is less than two times the smallest standard deviation), you can assume the variance in each group is the same. With experience, you will be able to "eyeball" this difference.

Homogeneity of Variances Rule of Thumb: if the largest standard deviation is less than two times the smallest standard deviation, the variance in each group is probably equal.

In the following example, four treatments are evaluated as the effect of the treatment on the height of plants.

Using the calculated standard deviations and the Homogeneity of Variances rule of thumb, we see the largest standard deviation is greater than two times the smallest standard deviation, and can conclude that the variances are probably unequal.

	Trt 1	Trt 2	Trt 3	Trt 4
	29	30	29	27
	34	32	30	28
	33	33	31	29
	34	34	34	33
	32	34	38	30
mean	32,4	32,6	32,4	29,4
SD	2,07	1,67	3,65	2,30

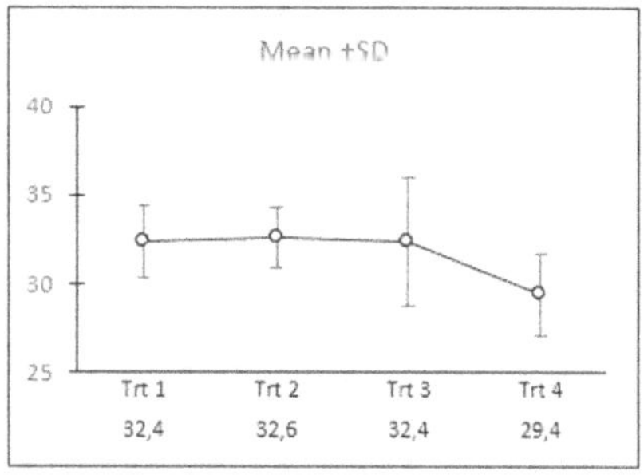

Homogeneity of variances - rule of thumb:	SD -greater	3,65
	SD - lesser	1,67
	Ratio	2,18
	Variances are:	**Unequal**

Figure 9.3: Comparing treatment variability using the Homogeneity of Variances Rule of Thumb.

Standard Error of the Mean (SE or SEM)

Standard Error of the Mean is derived from the Standard Deviation and quantifies the precision of the calculated Mean relative to the true population Mean, and considers the variability (Standard Deviation, SD) and the sample size (n) - factors which determine how close the sample Mean will be relative to the population Mean.

As for Standard Deviation, Standard Error of the Mean is expressed in the same units as the data, and is calculated as the Standard Deviation divided by the square root of the sample size:

$$SE = SD \,/\, SQRT(n)$$

It can be seen from the above equation that the Standard Error is always smaller than the Standard Deviation of a dataset and, in contrast to the Standard Deviation, the Standard Error always gets smaller as your sample size gets larger.

For this reason, Standard Error of the Mean is sometimes misused for data description, and it is generally recommended that 95% Confidence Intervals be used instead.

The 95% Confidence Interval (95% CI)

The 95% Confidence Interval is derived from the Standard Error of the Mean and quantifies the precision of the calculated Mean, such that the Confidence Interval contains the true population Mean with 95% certainty.

The 95% Confidence Interval of a Mean is calculated as the Standard Error of the Mean multiplied by a constant from the t-distribution for the given sample size (n-1 degrees of freedom, DF):

$$95\%CI = SEM * t(n\text{-}1)$$

While 95% Confidence intervals are easily calculated, it is occasionally useful to be able to give a rough estimate of these Intervals using the Confidence Interval Rule of Thumb:

Confidence Interval Rule of Thumb I: for large sample sizes, 95% Confidence Interval may be estimated from the Mean plus or minus two Standard Errors of the Mean.

The above is valid for large sample sizes, for smaller sample sizes the t-distribution provides a correction factor.

Referring to the t-distribution (Figure 9.4) we can see that t-values are close to two for datasets greater than twenty while for datasets less than 10 the t-value increases rapidly.

df	*p= .05 2-tailed*	df	*p= .05 2-tailed*
1	12.71	**13**	2.16
2	4.30	**14**	2.14
3	3.18	**15**	2.13
4	2.78	**16**	2.12
5	2.57	**17**	2.11
6	2.45	**18**	2.10
7	2.36	**19**	2.09
8	2.31	**20**	2.09
9	2.26	**30**	2.04
10	2.23	**50**	2.01
11	2.20	**100**	1.98
12	2.18	∞	1.96

Figure 9.4: t-distributions for given sample sizes (n-1 degrees of freedom, DF) for the calculation of 95% Confidence Intervals.

Plotting and interpreting SD, SEM and 95% CI Error bars

When plotting data with Error bars or creating tables with +/- values, SD, SEM, and 95%CI are often applied and interpreted incorrectly. Almost as frequently, Error bars are presented without any mention of what they indicate.

Let us begin by calculating the Mean Standard Deviation, Standard Error of the Mean and 95% Confidence Interval of a set of data. In the following example (Figure 9.5), the Mean is illustrated by use of a bar graph, to which the respective Error bars are added.

As we can see, the Standard Deviation Error bars are broader (further from the Mean) than those of the Standard Error of the Mean.

This is because the Standard Error of the Mean is calculated as the Standard Deviation divided by the square root of the sample size.

For this reason, Standard Errors of the Mean are sometimes presented as Error bars without any indication of their identity, with the intent of making variability seem as small as possible.

Even worse, this practice may mislead those who erroneously interpret non-overlapping Error bars as indicative of significant differences between the Means.

Standard Deviations should only ever be used to express the variation in our data and showing the variation in our data as a scatter plot usually provides more information then expressing the variation as Standard Deviation Error bars.

Standard Error of the Mean (SEM) Error bars may be used to express how accurately your data define the Mean, but if your objective is to compare Means, 95% Confidence Interval Error bars are more useful.

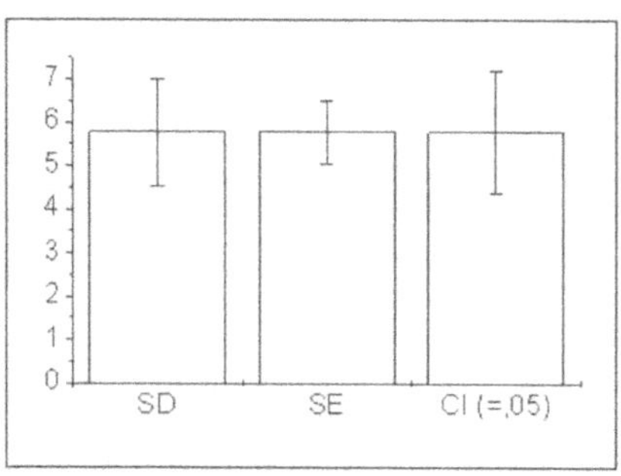

Figure 9.5: Bar graph for the Mean of a dataset showing Error bars for Standard Deviation (SD), Standard Error of the Mean (SE) and 95% Confidence Interval (CI), respectively.

In addition, Confidence Intervals may be used to provide an initial estimate of statistical significance, and Confidence Intervals are often easier to interpret than statements about statistical significance.

As an initial indication, if Confidence Intervals do NOT overlap, the Means are probably significantly different (at the $p < 0.05$ level). This may be confirmed as required using a valid statistical test (*e.g.* a two-sample t-test).

Confidence Interval Rule of Thumb II: if Confidence Interval Error bars do not overlap, the differences between the Means are probably significant; if Confidence Interval Error bars overlap by more than 25%, the differences between the Means are probably not significant.

It is important to be aware that the opposite is not necessarily true (a mistake made by students and scientists alike): if Confidence Intervals DO overlap, there may be a statistically significant difference between the Means, and it is thus NOT possible to deduce that the difference between the Means is NOT significant.

However, a statistical rule of thumb states that if Confidence Interval Error bars overlap by more than 25%, the Means are probably not significantly different at the $p < 0.05$ level) - to confirm this one should conduct a two-sample t-test and calculate the p-value.

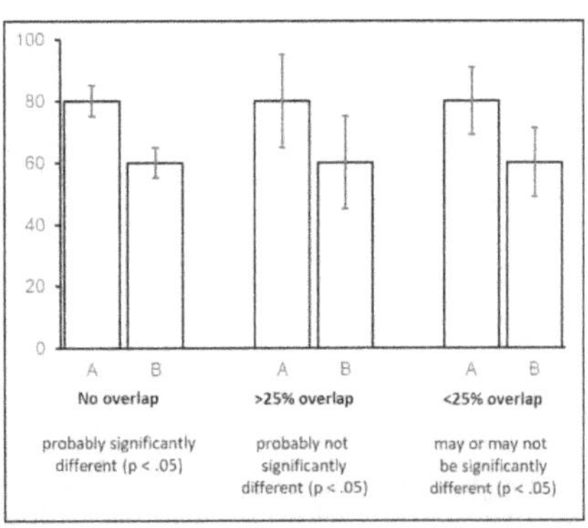

Figure 9.6: Use of 95% Confidence Interval (95% CI) Error bars to estimate the significance of differences between Means.

In conclusion: when used correctly, Standard Deviation, Standard Error of the Mean and 95% Confidence Intervals are useful descriptive statistics allowing us to critically evaluate our experimental data for variability (SD), proximity of the calculated Mean to the population Mean (SE) and may be used as a visual estimate of whether the differences between Means are significant (95%CI).

The principal advantage (and potential disadvantage) of using 95% Confidence Intervals to estimate the significance of differences between Means is that these calculations may be easily performed and interpreted without access to - or an understanding of - advanced statistical techniques.

10

IDENTIFICATION OF OUTLIERS: Q-TEST, IQR AND SD

In one of the first chapters, we began with the observation that a herbicide safener might reduce insecticide phytotoxicity, from which a research hypothesis was formulated.

To experimentally test the hypothesis we used significance levels, power, and effect to design a randomized complete block experiment in which plants were treated with insecticides in the absence or presence of safeners and evaluated for differences in phytotoxicity between treatments.

The experimental parameters were defined under the assumption that our data would be derived from an approximately normal data set.

We learned that for biological data, the assumption of normality is generally a valid approximation, as the normal distribution of biological data is surprisingly ubiquitous.

With the results of our evaluation in hand, the next step is to critically evaluate the experimental data by identifying and possibly rejecting outliers. Before proceeding, we should ensure that all values are correctly entered, and correct if necessary.

We then take experimental problems into account - if an apparent outlier derives from an experimental mistake or malfunction, it may be justified to exclude the value without performing additional calculations.

It should be noted that in addition to detecting outliers with the intention of eliminating them from subsequent statistical analyses, there is also the option of detecting outliers to identify datasets that qualify for retesting, or to identify outliers with the intention of retaining them for analysis using so-called robust statistical techniques.

The need to identify and remove outliers is reduced when the data is to be analyzed using "robust" methods of statistical analysis, such as nonparametric tests which compare the distribution of ranks.

Here the largest value will have the highest rank irrespective of how large that value is. If the data is to be statistically analyzed using parametric tests (which assume a Gaussian distribution, or Normalcy) the need to eliminate outliers is far greater.

Q-test for Outliers

The simplest method to identifying outliers in smaller datasets (n = 3-10) typical for biological trials is to use the Q-test published in 1951 by Robert Dean and Wilfred Dixon, for which we again assume Normalcy of our data (the Q-test prescribes that the data - excluding the possible outlier - must be Normally distributed).

We begin by arranging the data in order of increasing value and calculate Q as the Gap divided by the Range, where the Gap is the difference between the outlier and the closest number to it and the Range is the difference between the lowest and the highest value.

$$Q = GAP / RANGE$$

The null hypothesis for the Q-test is: there are no outliers in the dataset. If the value of Q is higher than the associated table value corresponding to the sample size and Confidence level (typically the 95% Confidence level, alpha = 0.05), the outlier may be rejected.

In the following experimental dataset, we wish to determine whether the value 77 is an outlier at the 95% Confidence level:

84 85
77 82 84 85 86 87

.....GAP......
....................RANGE.........

After sorting the data into ascending order, we can use the following formula to calculate the Q statistic by dividing the gap (5) by the range (10):

$$Q = GAP / RANGE = 5 / 10 = 0.5$$

From the schematized table of critical values for Q (below) we can quickly determine that our calculated Q statistic of 0.5 is less than the critical two-sided (*i.e.* determining for outliers at both extremes of the dataset) table value of 0.53 for a sample size of 8 at the most commonly used 95% Confidence level, and the data value 77 may thus NOT be rejected:

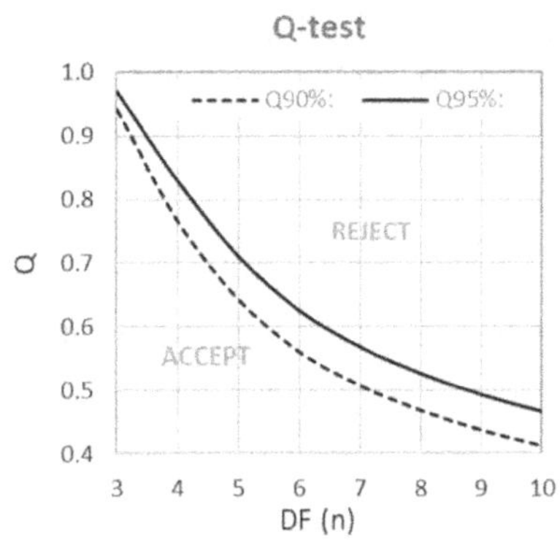

Figure 10.1: Outliers - schematized two-sided table for critical values of Q for 90% and 95% Confidence levels.

From the schematized table of critical values for Q (above) we can determine that our calculated Q statistic of 0.5 is less than the critical two-sided (*i.e.* determining for outliers at both extremes of the dataset) table value of 0.53 for a sample size of 8 at the most commonly used 95% Confidence level, and the data value 77 may thus NOT be rejected.

The Q-test for outliers is a mathematically simple procedure but does require that we take the time to consider the results logically: for example, performing the Q-test on the values 82.24; 82.25 and 82.25 returns the conclusion that 82.24 is an outlier! Clearly, statistical significance does not always equal practical significance.

An alternative method in common use for larger datasets is Grubb's test for outliers, in which outliers are identified as the difference between the outlier and the Mean, divided by the Standard Deviation. This ratio is then compared to a table of critical values.

The advantage of the Q-test relative to Grubb's test is that calculations may easily be made by hand, the disadvantage is that it is only valid for data sets up to n=10. However, for most experimental biological applications the Q-test is both appropriate and convenient.

. . .

Interquartile Range (IQR) Test for Outliers

Box plots are a useful tool for identifying outliers in a dataset. Outliers are data points that are significantly different from the rest of the data and can skew the results of statistical analyses. Box plots display the distribution of data in a way that makes it easy to identify outliers.

The box in the plot represents the interquartile range (IQR), which is the range between the first and third quartiles. Any data points that fall outside of the whiskers of the box plot are considered outliers. The whiskers extend to the minimum and maximum values of the data, but any data points beyond 1.5 times the IQR from the edge of the box are considered outliers and are plotted as individual points.

The Interquartile Range (IQR) is a statistical measure that helps identify outliers in a dataset. It is determined by the range between the first quartile (Q1) and the third quartile (Q3) of the data.

To use the IQR for outlier identification, we calculate Q1 and Q3 and then find the range of values that lie within 1.5 times the IQR below Q1 or above Q3. Any data point that falls outside this range is considered a potential outlier.

Let's consider an example with two herbicide treatments. The mean height for treatment 1 is 70cm, and the mean height for treatment 2 is 38cm.

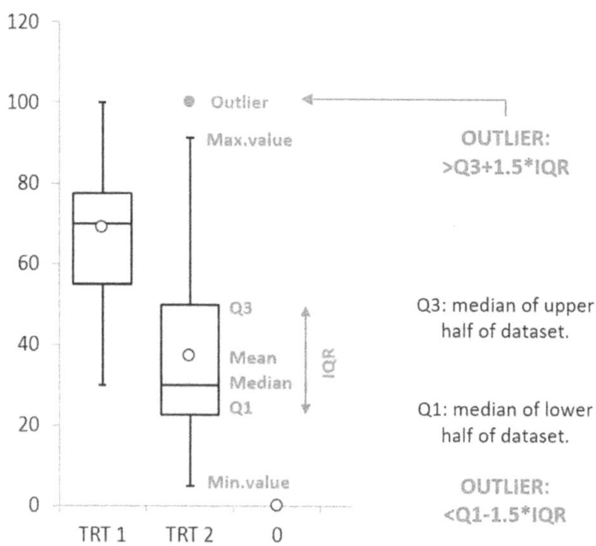

Figure 10.2: The Interquartile Range (IQR) to identify outliers in a dataset.

To determine outliers using the IQR, we start by creating a box plot. The box plot displays the distribution of data, with a box representing the IQR (the range between the first quartile, Q1 and the third quartile, Q3).

A line inside the box indicating the median, a circle to show the mean and "whiskers" extending from the box to show the range of non-outlier values (the range of values that lie within 1.5 times the IQR below Q1 or above Q3). Any data points beyond the whiskers are considered potential outliers.

By observing the box plot, we see that the value of 100cm in treatment 2 falls outside the whisker range, indicating it as a potential outlier.

SD Test for Outliers

An even simpler (but less accepted) method to detect and eliminate outliers is by using the standard deviation of the data distribution.

We may assume a value to be an outlier if it is more than 2 or 3 times the standard deviation of the distribution.

This method is based on the Rule of Thumb for Normal Distribution: The approximate percentage of values falling within one, two, and three standard deviations of the mean in a normal distribution is 68%, 95%, and 99.7%, respectively.

Data presentation for researchers and decision makers

For **researchers** evaluating the data, the box plot is a valuable tool to gain insights into the quality of the data and identify important characteristics. The box plot provides a visual summary of the data distribution, displaying the median, quartiles, and potential outliers.

By examining a **box plot**, researchers can quickly assess measures like skewness (indicating a departure from symmetry) and identify any outliers that might be influential or require further investigation. This helps researchers understand the overall pattern and variability in the data, aiding in data interpretation and potential adjustments to analysis methods.

When presenting data to **decision-makers** who may not have a statistical background, using a bar chart with error bars can be a helpful visual tool.

This type of chart shows the means of different groups or treatments, along with the 95% Confidence Interval (CI) represented by the error bars. The CI provides a range of values around the mean that we are reasonably confident the true population mean falls within.

Decision-makers often assume that if the error bars of two groups or treatments do not overlap, then the means are significantly different. While this assumption is not always accurate, the CI at least gives decision-makers a probability of making the right decision based on the available data. It's important to note that statistical significance should be properly tested using appropriate hypothesis tests (next section) rather than relying solely on visual cues.

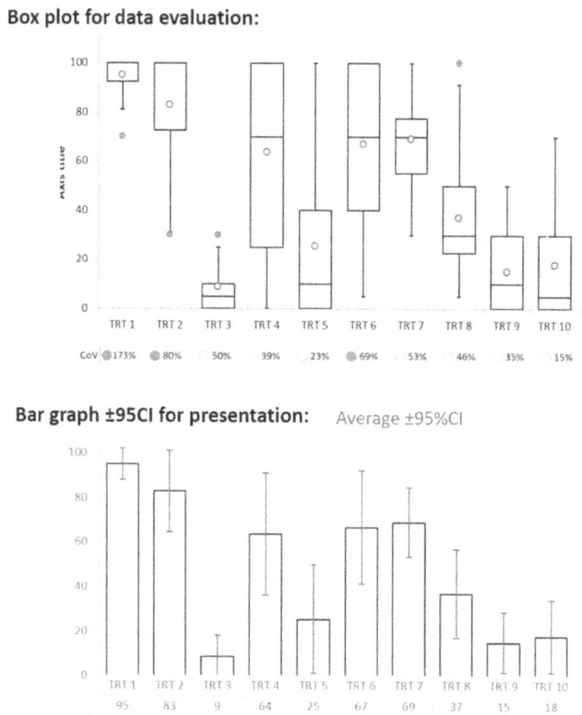

Figure 10.3: Data presentation for researchers and decision makers

By using **bar charts with 95% CI error bars** when presenting data to decision-makers and box plots during data evaluation, both parties can gain valuable insights.

Decision-makers obtain an understanding of potential differences between means based on the visual cues, while researchers can examine the data's characteristics more thoroughly and identify outliers or other important features that may impact the statistical analysis.

Ultimately, the combined use of these two visual tools enhances both the communication of findings and the researcher's ability to draw meaningful conclusions from the data.

11

CHI-SQUARE TEST: GOODNESS OF FIT FOR SAMPLE VALIDATION

The chi-squared distribution or $\chi 2$-distribution, developed by English mathematician and statistician Karl Pearson in 1900 is a widely used probability distribution in inferential statistics (Section IV) and is a component of the t-distribution, f-distribution, the Analysis of Variance (Chapter 17) and regression analysis (Section V).

Chi-squared tests for the goodness-of-fit (within-sample deviations)

The chi-squared distribution is also used in chi-squared tests for the goodness of fit of an observed distribution - a component of experimental validation - by comparing expected and observed values.

In a typical validation study example, a within-sample comparison for individual observed values of bacterial counts on Petri dishes against average observed values is required to conform to an acceptance criterium, for example, within-sample deviation of $\chi 2$ must conform to $p>0.05$ (see example below).

If the chi-square test indicates that the test result exceeds the stipulated criteria limit (we speak of the chi-square test failing to support the Null hypothesis: there is no difference between the counts) possible outliers may be removed from the evaluation.

No more than 20% of the counts (one in five) may be removed. If the Chi-square test fails after the removal of possible outliers - *i.e.* if $p<0.05$ - the sample is considered invalid and must be rejected.

In the following example, bacterial counts of a microbial biopesticide are performed on five Petri dishes. The Null hypothesis is that there is no difference between the observed counts (O), relative to the average (and thus expected, E) count:

Plate	1	2	3	4	5
Count (O)	60	61	66	60	66
Average (E)	62.6				
(O-E)^2/E	0.11	0.04	0.18	0.11	0.18
Σ(O-E)^2/E	**0.63**				

The deviation for each count relative to the average (E) is calculated as:

$$(O\text{-}E)^\wedge 2/E$$

...while the $\chi 2$ value (the sum of deviations for each count relative to the average (E) is calculated as:

$$\chi 2 = \Sigma(O\text{-}E)^\wedge 2/E$$

The calculated probability (P; $\chi 2$) - the likelihood that the within-sample deviation between expected (E) and observed (O) counts is due to chance - can be compared to the table values in Figure 11.1 for *n-1* degrees of freedom (df).

If the $\chi 2$ probability lies above the 5% significance curve ($p<0.05$; Figure 11.1) the probability that the deviations are due to chance are small - the sample is considered invalid and must be rejected.

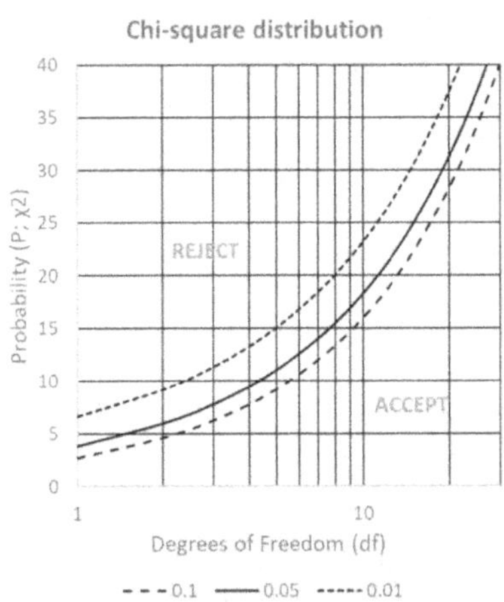

Figure 11.1: $\chi 2$ distributions for 2-20df; for 1%, 5% and 10% significance.

Conversely, if the $\chi2$ probability falls below the 5% significance curve ($p>0.05$) the probability that the deviations are due to chance is great - the sample conforms to the acceptance criteria and can be accepted.

For our example, the calculated within-sample comparison values for individual observed counts against average observed counts ($\chi2=0.63$) are less than the critical value of 9.49 for 4df at 5% significance (Figure 11.1) and the sample thus conforms to the acceptance criteria and can be accepted.

12

CORRECTED EFFICACY FOR UNTREATED CONTROL MORTALITY

When evaluating the efficacy of insecticides, the assumption is made that the observed responses are due to the treatments only. However, in some cases the responses may occur independently of the treatments at zero dose (untreated control), for example through death of the untreated targets from natural causes.

In such cases it is possible to compensate for this by calculating corrected efficacy.

If the control mortality is above 20%, the results must be discarded. When control mortality is greater than 5% but less than 20%, the observed mortality must be corrected using the appropriate formula. If the control mortality is below 5%, it can be ignored, and no correction is necessary.

Corrected efficacy is a statistical approach that helps researchers accurately assess the effectiveness of a pesticide or treatment. By subtracting the control mortality from the treatment mortality, we isolate the mortality directly caused by the treatment.

This allows us to focus on the treatment effect specifically, separate from other factors. The purpose is to eliminate confounding factors and determine the true impact of the treatment in controlling pests or diseases.

Corrected efficacy can be calculated by appropriate methods, depending on whether the population is uniform (if there were no significant differences between the means in ANOVA *before* the treatment) or non-uniform, and whether the observation is based on infestation (live individuals) or mortality (dead individuals).

Uniform pest or disease infestation before treatment i.e., no significant differences between the means (ANOVA) - correction for natural mortality in untreated control plots

Non-uniform pest or disease infestation before treatment i.e., significant differences between the means (ANOVA) - correction for natural mortality and non-uniformity of infestation in untreated control plots

---

Abbott's method, Henderson-Tilton's method, Schneider-Orelli's method, and Sun-Shepard's method are all statistical approaches commonly used in agricultural and biological research to analyze and compare pesticide or treatment effects on pests or diseases. Here is a brief overview of each method:

Methods for corrected efficacy for infestation (live counts) data:

Abbott's method: Calculates efficacy based on treatment mortality as a percentage of the untreated control (Uniform pest or disease infestation before treatment).

Henderson-Tilton's method: Calculates efficacy based on change in infestation before and after treatment. It accounts for the natural or background mortality observed in the control group and allows for a more accurate estimation of the mortality caused by the pesticide treatment (Non-uniform pest or disease infestation before treatment).

Uniform disease or pest infestation before treatment	**Abbott's method:** $$Corrected\ \% = (1 - \frac{Population\ in\ treated\ plot\ after\ treatment}{Population\ in\ control\ plot\ after\ treatment}) * 100$$
Non-uniform disease or pest infestation before treatment	**Henderson-Tilton's method:** $$Corrected\ \% = (1 - \frac{n\ in\ Co\ before\ treatment * n\ in\ T\ after\ treatment}{n\ in\ Co\ after\ treatment * n\ in\ T\ before\ treatment}) * 100$$

Figure 12.1 : Abbot's method (uniform infestation) and Henderson-Tilton's method (non-uniform infestation) for corrected efficacy, infestation (alive)

Methods for corrected efficacy for mortality evaluation:

Schneider-Orelli's method: Calculates efficacy based on treatment mortality as a percentage of the untreated control (Uniform pest or disease infestation before treatment).

Sun-Shepard's method: Calculates efficacy based on change in control mortality before and after treatment (Non-uniform pest or disease infestation before treatment).

Uniform disease or pest infestation before treatment	**Schneider-Orelli's method:** $Corrected\ \% = (\frac{\%\ mortality\ in\ T - \%\ mortality\ in\ Co}{100 - \%\ mortality\ in\ Co}) * 100$
Non-uniform disease or pest infestation before treatment	**Sun-Shepard's method:** $Corrected\ \% = (\frac{\%\ mortality\ in\ T - \%\ change\ in\ Co}{100 - \%\ change\ in\ Co}) * 100$ $\%\ change\ in\ Co = \frac{n\ in\ Co\ after\ treatment - n\ in\ Co\ before\ treatment}{n\ in\ Co\ before\ treatment} * 100$

Figure 12.2: Schneider-Orelli's method (uniform infestation) and Sun-Shepard's method (non-uniform infestation) for corrected efficacy, mortality.

These methods may be simplified and summarized as follows:

	Uniform population	Non-uniform population
Infestation (live)	Abbot's % correction: $1 - \frac{nTa}{nCa} * 100$	Henderson-Tilton % correction: $1 - \frac{nCb * nTa}{nCa * nTb} * 100$
Mortality (dead)	Scheider-Orelli % correction: $1 - \frac{m\%T - m\%C}{100 - m\%C} * 100$	Sun-Shepard % correction: $1 - \frac{m\%T - \Delta\%C}{100 - \Delta\%C} * 100$

nT(C)a = number of insects in Treatment (Control) after treatment, and nT(C)b = number of insects in Treatment (Control) before treatment
m%T(C) = mortality % in Treatment (Control) and Δ%C = change % in Control.

Figure 12.3: Summary of methods for calculating corrected efficacy.

The following examples from crop protection trials show how the formulas translate into actual calculations and interpretations.

	Uniform population	Non-uniform population
Infestation (live)	Example: Abbot % correction In an experimental plot, 40 insects are observed in the control plot while only 30 insects were observed in the treated plot. The corrected efficacy is calculated as: $Corrected\ \% = \left(1 - \frac{30}{40}\right) * 100 = 25\%$	Example: Henderson-Tilton % correction After treatment, 40 insects are observed in the control vs only 30 insects in the treated plot, but 8 and 6 insects respectively were found before treatment: $Corrected\ \% = 1 - \frac{8 \cdot 30}{40 \cdot 6} * 100 = 0\%$
Mortality (dead)	Example: Scheider-Orelli % correction In an experimental plot, 10% mortality is observed in the control plot while 50% mortality is observed after treatment. The corrected efficacy is calculated as: $Corrected\ \% = \left(\frac{50-10}{100-10}\right) * 100 = 44\%$	Example: Sun-Shepard % correction In an experimental plot, a 25% change in mortality (from 8 to 10%) is observed in the control plot while 50% mortality is observed after treatment. The corrected efficacy is: $Corrected\ \% = \left(\frac{50-25}{100-25}\right) * 100 = 33\%$

Figure 12.4: Examples, corrected efficacy calculations.

IV

INFERENTIAL STATISTICS - ACCEPTING OR REJECTING THE HYPOTHESIS

13

DATA SCALES (DATA TYPES)

In the previous chapters, *Descriptive statistics* were shown to provide a summary of our dependent variable data in the form of measures of central tendency (mean) and measures of variability (variance, standard deviation, standard error of the mean, 95% confidence intervals, etc.).

In contrast to Descriptive statistics on sample data, *Inferential statistics* move beyond the immediate data to infer the significance of observed differences between treatments, and thus whether our data allows us to reject the null hypothesis.

Hypothesis testing

Hypothesis testing is a means of drawing conclusions about a population mean, and a range of tests are available to determine whether a hypothesis about the mean is true or not.

The choice of test is determined by the scale (type) and distribution of the data. The following section provides a brief overview of data scales or types.

Types of Data

Most experimental biological data are in the form of *measurement data*, or data expressed using an interval or ratio scale. The following provides a brief hierarchical overview of the four principal data scales, or types.

. . .

Nominal and Ordinal data

Nominal (non-numeric) and Ordinal (categorical numeric) data are rarely encountered in biological testing and comprise categorical or discrete variables and are typically described by two or more values (categories). Due to their non-numeric nature, it is possible to differentiate, but not to order or rank Nominal (such as "male" / "female") data.

Category	Type	Data	Description	Power	Variable	Descriptive statistics	Inferential statistics
Measurement	Numerical	Discrete or continuous	Amount of Difference; absolut zero	Increasing information	RATIO	Mean, SD, SEM, 95%CI	Parametric or non-parametric
			Amount of Difference		INTERVAL	Median, Min /Max, Range	
Rank / Count	Categorical	Discrete	Direction of Difference		ORDINAL	Mode, Counts, Frequency	Non-parametric
			Difference		NOMINAL		

Figure 13.1: Hierarchy of dependent variable data scales (types) and descriptions, including a summary of appropriate descriptive and inferential statistical analysis methods.

In contrast, it is possible to rank Ordinal data (such as shortest to tallest plant), providing an additional aspect of direction to the simple binary differentiation of Nominal data.

However, ordinal data categories do not have an equivalent distance between them, so a relative magnitude of the difference cannot be determined.

Nominal and Ordinal data may be summarized using descriptive statistics (mode, count, frequency) and hypotheses may be tested using *nonparametric* inferential statistics which do not assume a Normal or Gaussian distribution (see next chapter).

As Nominal and Ordinal data do not indicate the magnitude of the difference, they are referred to as *rank* or *count data.*

Interval and Ratio data

In contrast to Nominal and Ordinal data, Interval and Ratio data allow us to determine the magnitude of the difference and are thus referred to as measurement data.

Interval and Ratio data comprise continuous (as opposed to discrete) numeric variables and may contain any value with a finite or infinite interval, such as plant height, insect mass, disease incidence or -severity.

Data on an Interval scale do not contain a true zero (for example, temperature measured in degrees Centigrade or Fahrenheit), while data on the Ratio scale (such as weight or height) provide a true or absolute zero, from which we, for example, may determine that a treated plant is double as tall as an untreated control plant.

Interval and ratio data may be summarized using descriptive statistics (median, minimum/maximum, range) for interval data, or (mean, standard deviation, standard error of the mean and 95% confidence interval) for both interval and ratio data.

Hypotheses may be tested using *parametric* inferential statistics (which assume a Normal or Gaussian distribution) - the topic of the next chapter.

14

NONPARAMETRIC OR PARAMETRIC STATISTICAL TESTS

Inferential statistics enable us to test a hypothesis and draw conclusions, or inferences regarding a population through extrapolation from our experimental data sample.

Our choice of statistical method for hypothesis testing is based on whether the experimental data is normally distributed, and on the scale of the data.

Assumption of Normal (Gaussian) distribution

Parametric tests such as the *t*-test and ANOVA are generally more powerful than non-parametric tests and are more likely to detect a significant effect when one indeed exists. They assume that the samples follow a normal (Gaussian) distribution. A number of tests of normality are available, such as the **Shapiro–Wilk and Kolmogorov–Smirnov tests**, but these require statistical insight or access to statistical software packages.

A simpler graphical approximation to evaluate normality is to visually determine how closely a histogram of the sample resembles a "bell-shaped" normal probability curve (Figure 14.1).

If the shape of the histogram looks approximately symmetrical and bell-shaped, the assumption of normalcy can be considered reasonably met. A box plot (inset, Figure 14.1) also provides an overview of data distribution and can identify non-normal distributions.

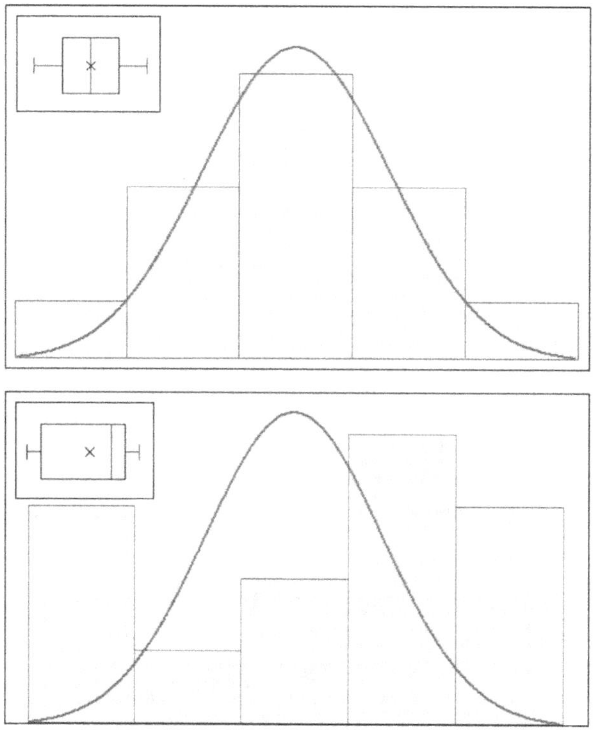

Figure 14.1: Histograms of sample data overlaid with a normal probability curve and box-plot (inset) identifying normal (top) and non-normal (bottom) data distributions.

A general rule of thumb is that the distribution of the mean approaches a normal distribution as sample size increases, and that to compare means your sample size should be at least 30.

If our data is not normally distributed, we have three options: to transform our data (*e.g.* log transformation) to normalize the distribution, to use a nonparametric test to compare the means, or simply to use the *t*-test despite the non-normalcy of our data (the *t*-test is considered to be robust to violations of assumptions of normalcy).

The most common transformations used for non-normal data include logarithmic, square root, and reciprocal transformations. These transformations apply mathematical functions to the data, altering the scale and shape of the distribution to better approximate normality, allowing for more reliable analysis using parametric tests.

As can be observed in the graphs, two common Transformations to normalize data (Square root transformation (Y=SQRT(X), Log transformation (Y=log(X)) reduce the impact of extreme values and making the distribution more symmetric:

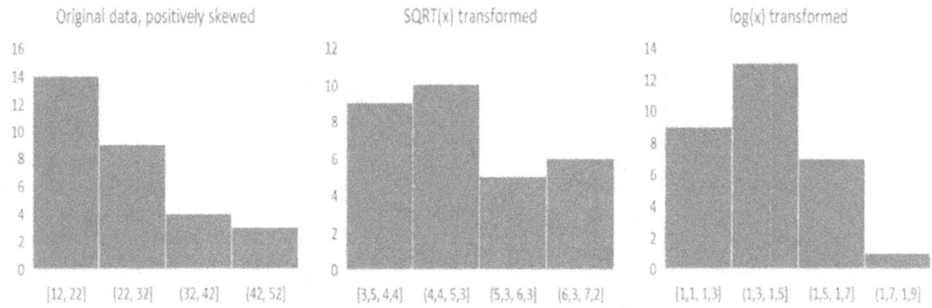

Figure 14.2: Transformation reduces the impact of extreme values and making the distribution more symmetric.

A further requirement for parametric tests is that of **homogeneity of variance**, meaning that the variances within the different treatments should be approximately equal. If one treatment is significantly more or less variable than the others, the analysis of variance (ANOVA) may not yield accurate results.

To determine if this requirement is met, and If you have access to a statistical package, you can use Bartlett's test or Levene's test to assess the equality of variances across multiple groups or samples. Bartlett's test performs well with larger sample sizes, and is more sensitive to deviations from normality, while Levene's test is suitable for both small and large sample sizes and is less sensitive and more robust in the presence of non-normality.

The F-test is commonly used in ANOVA to compare variances between groups assuming normality and equal variances.

Alternatively, a box and whisker graph can be examined, and you can critically evaluate your data by comparing the variability of samples using the Homogeneity of Variances Rule of Thumb:

Homogeneity of Variances Rule of Thumb: if the largest variance is no more than 4 times the small variance (i.e., if the largest standard deviation is less than two times the smallest standard deviation), you can assume the variance in each group is the same.

Outlier removal and Transformations to reduce variability

- Outlier removal eliminates extreme values that deviate significantly from the overall pattern of the data, reducing their impact and making the dataset more reliable. Q-test, IQR test, SD test
- Transformations to normalize data such as logarithmic transformation can help reduce the variability of skewed distributions or data with a wide range of values. Logarithmic transformations compress larger values more than smaller values, which can reduce the spread of the data and make it more symmetrical.

Parametric and non-parametric tests

For normally distributed data, standard parametric tests such as the T-test and ANOVA tests are typically used, while nonparametric tests are appropriate if the data does not follow the normal distribution.

Parametric tests assume a Normal or Gaussian distribution of Measurement data at the Interval or Ratio scales (see previous chapter), while nonparametric do not - although they are subject to sample size requirements (see below).

In addition to non-Gaussian Measurement data, Nonparametric tests are used for Categorical data at the Nominal or Ordinal scales.

As Parametric tests are more powerful than non-parametric tests and more likely to detect a significant effect when one indeed exists, many biologists tend to favor parametric tests rather than nonparametric tests, as any non-conformity to the prerequisites for parametric testing can often be circumvented through assumptions of normalcy, the identification and removal of outliers as well as data transformations.

In daily practice, it is thus usual for scientists to transform data from non-normal distributions, or to use parametric methods directly on datasets from non-normal distributions.

In cases where each treatment consists of fewer than 10 data values, as is common in crop protection studies, it is generally accepted that conducting tests for normal distribution becomes unreliable.

A biological rule of thumb is that the small data sets commonly found in lab and greenhouse trials may be assumed to be normally distributed and analyzed using standard parametric tests.

Therefore, neither data transformation nor the use of nonparametric tests would yield significant benefits. As a general rule of thumb in biology, small data sets commonly encountered in laboratory and greenhouse trials can be assumed to be Normally distributed, and standard parametric tests can be employed for analysis.

Overview of essential parametric and non-parametric methods

For hypothesis testing, specific parametric and nonparametric tests are available to evaluate different experimental datasets.

	Measurement (Gaussian) data	Categorical or Measurement (non-Gaussian) data
	Parametric	**Nonparametric**
Describe one group	Mean SD; SEM; 95%CI	Median Range
Compare one group to a hypothetical value	1-sample t-test	Wilcoxon test (or Chi-square for two possible outcomes)
Compare two unpaired groups	2-sample unpaired t-test	Mann-Whitney test
Compare two paired groups	2-sample paired t-test	Wilcoxon test
Compare three or more unmatched groups	One-Way ANOVA	Kruskal-Wallis test ((or Chi-square for two possible outcomes)
Predict value from measured values	*Linear or nonlinear regression*	*Nonparametric regression*

Figure 14.3: Overview of essential statistical analysis methods.

Among these, the most commonly used include tests to compare a single group of data to a hypothetical value, to compare two paired or unpaired groups, or to compare three or more groups, as well as the prediction of values from previously measured values.

NOTE: the information presented here comprises approximations and rules of thumb which are commonly used in designing and analyzing non-critical trials only. For critical experiments, nonparametric tests should be used if the data does not follow the normal distribution (before or after transformation) or sample size requirements for parametric tests - always seek the advice of a qualified statistician!

Microsoft Excel versus Statistics packages

This would be an appropriate time to spend a minute or two considering software packages for statistical data analysis. Microsoft Excel is perhaps the most commonly used data software in biological research and is often used (and misused) for Statistical Data Analysis.

Excel is first and foremost a spreadsheet with added data analysis modules, and it is important to understand that the software has limitations relative to professional statistic packages.

Excel is however excellent for data entry and data management and is accessible and sufficiently applicable for "quick-and-dirty" internal descriptive analyses and initial hypothesis testing carried out in research laboratories.

For more demanding applications (*e.g.* external reports and scientific papers) statisticians advise that Excel should only be used for data preparation, and this data should then be transferred to professional statistics packages for analysis. These results can then be reported directly or moved back to Excel for graphing and presentation purposes.

In Crop Protection Research, two of the most used data packages for trial planning and statistical data analysis are the commercial ARM (Agricultural Research Manager) package, and the open-source statistical package, R.

Both have relatively steep learning curves, but once they have been mastered, they become indispensable. For researchers with limited scientific knowledge, the GraphPad suite of statistical packages provides real-time guidance during data analysis.

We will obtain insight into the inferential methods commonly used to determine whether the means of the measured variables are equal (the null hypothesis) or whether they are significantly different (t-tests and ANOVA) in the following chapters.

15

TWO SAMPLE MEANS COMPARISON: THE T-TEST (PART I)

For biological experiments, the objective of most trials is to determine whether the means of the measured variables are equal (the null hypothesis) or whether they are significantly different.

The *t*-tests are a group of inferential parametric methods used to determine if two samples (or a sample and a theoretical mean) have the same mean (null hypothesis), or if there is a significant difference between the means.

The *t*-tests use the *t*-statistic (also termed *t*-value or *t*-ratio), *t*-distribution values and degrees of freedom to determine the difference probability between two data samples.

Different *t*-tests are used depending on whether a single sample is to be compared to a mean predicted by theory (one sample *t*-test), whether the samples are dependent (paired two-sample *t*-test) or independent (two-sample *t*-test) and whether the variances of the two samples are equal (two-sample *t*-test assuming equal variances) or not (two-sample *t*-test assuming unequal variances).

The one-sample *t*-test allows us to compare the mean of a single sample with an expected mean based on prior data (or a theoretical value), to determine if they are significantly different.

For example, if we - as part of an annual validation of our seed stocks - measured the mean weekly growth rate of ten seedlings from our in-house stock of spring barley seeds, we could compare this mean to last year's mean by performing a one-sample *t*-test to determine whether this year's mean growth rate was significantly different from last years recorded mean (the null hypothesis is that seedling vigor is unchanged, and that there will be no significant difference in weekly growth rate).

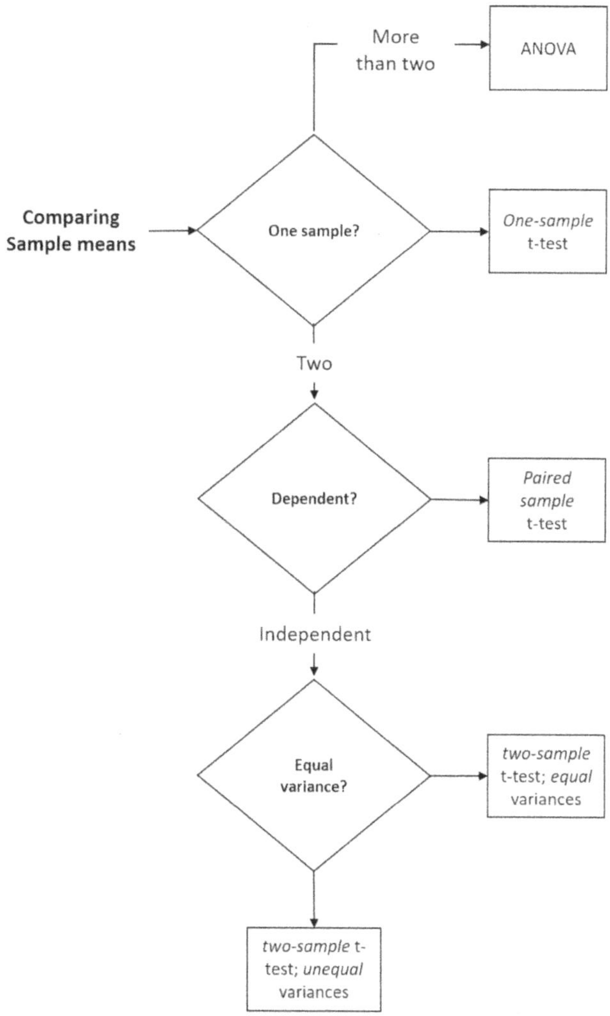

Figure 15.1: Summary of parametric inferential methods used to compare two sample means.

The paired *t*-test allows us to compare two sample means if each value within one sample can be paired with an associated value in the other, to determine if they are significantly different.

For example, if we measured the mean weekly growth rate of the same seedlings from our in-house stock of spring barley seeds over a period of consecutive weeks, we could compare this week's mean to last week's mean by performing a paired *t*-test to determine whether this week's mean growth rate is significantly different from last week's mean.

The (unpaired) two-sample *t*-test allows us to compare two sample means, to determine if they are significantly different.

For example, if we measured the mean weekly growth rate of herbicide-treated seedlings to untreated seedlings, we could use a two-sample *t*-test to determine whether the mean growth rate for the treated seedlings was significantly different from the untreated seedlings.

	One-sample *t*- test	Paired *t*- test	Two-sample *t*- test
	Single sample compared to predicted mean	*Two dependent samples*	*Two independent samples*
Sample variable:	Mean weekly growth rate *(current)*	Mean weekly growth rate *(this week)*	Mean weekly growth rate *(untreated)*
Compared variable:	Mean weekly growth rate *(last year)*	Mean weekly growth rate *(last week)*	Mean weekly growth rate *(treated)*

Figure 15.2: Example cases for one sample t-test for single data samples, paired two-sample t-test for dependent samples or two-sample t-test for independent samples.

The *t*-test assumes that the samples follow a normal (Gaussian) distribution.

A general rule of thumb is that the distribution of the mean approaches a normal distribution as sample size increases, and that to compare means your sample size should be at least 30.

For situations where a sample size of 30 is not feasible, the *t*-test was developed specifically to improve means testing results for small samples (typically less than 10 degrees of freedom).

Especially for samples with less than 5 degrees of freedom, the sharply increasing values of *t* (Figure 15.3) compensate for small sample size tendencies toward Type I errors (claiming a difference is significant when it is not).

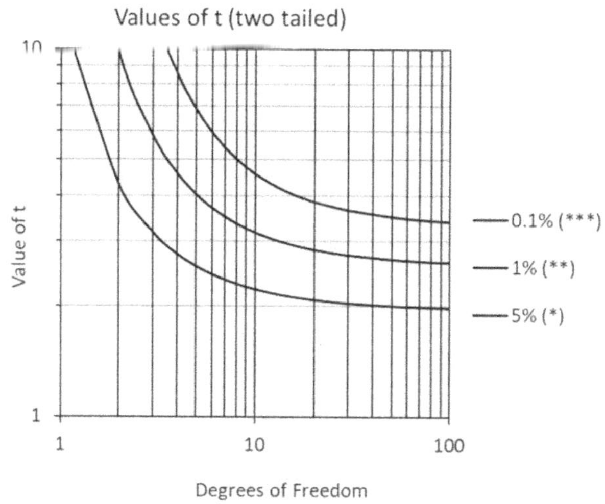

Figure 15.3: Values of t (two-tailed) for varying levels of significance.

Finally, the ***t*-test assumes that the data is sampled from populations that have equal variances**, even if their means are different.

As part of the (unpaired) two-sample *t*-test analysis, the *f*-test is used to determine if the two samples have equal variances - the topic of the next chapter.

16

TWO SAMPLE MEANS COMPARISONS: THE F-TEST

The (unpaired) two-sample *t*-test assumes that the data is sampled from populations that have equal variances, even if their means are different. As part of the (unpaired) two-sample *t*-test analysis, the *f*-test[1] is used to test if two samples have equal variances.

The *f*-test computes the ratio of the larger (greater) sample variance (square of the standard deviation, SD) relative to the smaller (lesser) variance. This *f*-ratio can be compared to one-tail critical *f*-table values ($\alpha = 0.05$) for n-1 degrees of freedom, df (Figure 16.1) to determine if the samples have the same values, *i.e.* the computed *f*-ratio is smaller than the critical *f*-value.

		df (greater)											
		1	2	3	4	5	6	7	8	9	10	20	∞
df (Lesser)	1	161	200	216	225	230	234	237	239	241	242	248	254
	2	18.5	19.0	19.2	19.3	19.3	19.3	19.4	19.4	19.4	19.4	19.4	19.5
	3	10.1	9.6	9.3	9.1	9.0	8.9	8.9	8.8	8.8	8.8	8.7	8.5
	4	7.7	6.9	6.6	6.4	6.3	6.2	6.1	6.0	6.0	6.0	5.8	5.6
	5	6.6	5.8	5.4	5.2	5.1	5.0	4.9	4.8	4.8	4.7	4.6	4.4
	6	5.6	5.1	4.8	4.5	4.4	4.3	4.2	4.2	4.1	4.1	3.9	3.7
	7	5.6	4.7	4.4	4.1	4.0	3.9	3.8	3.7	3.7	3.6	3.4	3.3
	8	5.3	4.5	4.1	3.8	3.7	3.6	3.5	3.4	3.4	3.4	3.2	2.9
	9	5.1	4.3	3.9	3.6	3.5	3.4	3.3	3.2	3.2	3.1	2.9	2.8
	10	5.0	4.1	3.7	3.5	3.3	3.2	3.1	3.1	3.0	3.0	2.8	2.5
	20	4.4	3.5	3.1	2.9	2.7	2.6	2.5	2.5	2.4	2.4	2.1	1.8
	∞	3.8	3.0	2.6	2.4	2.2	2.1	2.0	1.9	1.9	1.8	1.6	1.0

Figure 16.1: One-tailed f-table for $\alpha = 0.05$.

As $\alpha = 0.05$ corresponds to a 5% (or less) chance of discarding the null hypothesis (the samples have equal variances), if the computed value is less than the table value the null hypothesis may not be discarded *i.e.* the samples have equal variances.

If this is the case, the two sample means may be compared using the two-sample *t*-test for *equal* variances; alternatively, the two-sample *t*-test for *unequal* variances should be chosen.

Example: Comparing treated and untreated seedlings.

In the following example, the mean height of ten untreated seedlings is compared to the mean height of ten herbicidally treated seedlings.

The means of the two sample data sets are respectively 62 cm and 48 cm (Figure 16.2), and we wish to determine if the two samples have the same mean (null hypothesis), or if there is a significant difference between the means.

As we have two samples which are unpaired, an (unpaired) two-sample *t*-test will allow us to compare the two sample means, to determine if they are significantly different.

The next step is to use the *f*-test to determine whether the variances of the two samples are equal or not.

	Untreated	Treated
	63	53
	72	48
	54	43
	62	45
	57	55
	73	47
	53	41
	51	57
	61	53
	69	39
Mean:	62	48

Figure 16.2: Experimental sample data; mean height of ten untreated and herbicide-treated seedlings.

The ratio of the greater sample variance relative to the lesser variance (*f*-ratio) is 60.7/38.0 = 1.6, which is less than the table value of 3.2 for n-1 degrees of freedom *i.e.* 9df (Figure 16.3).

As the *f*-ratio is lower than the critical *f* value, we may conclude that the variances of the two samples are equal.

	Untreated	*Treated*
	63	53
	72	48
	54	43
	62	45
	57	55
	73	47
	53	41
	51	57
	61	53
	69	39
Mean:	*62*	*48*

	Untreated	*Treated*
Mean	61.58	48.12
Variance	60.68	37.96
Observations	10	10
df	9	9
F	1.6	
P(F<=f) one-tail	0.25	
F Critical one-tail	3.2	

df (Lesser)	df (greater) 1	2	3	4	5	6	7	8	9	10	20	∞
1	161	200	216	225	230	234	237	239	241	242	248	254
2	18.5	19.0	19.2	19.3	19.3	19.3	19.4	19.4	19.4	19.4	19.4	19.5
3	10.1	9.6	9.3	9.1	9.0	8.9	8.9	8.8	8.8	8.8	8.7	8.5
4	7.7	6.9	6.6	6.4	6.3	6.2	6.1	6.0	6.0	6.0	5.8	5.6
5	6.6	5.8	5.4	5.2	5.1	5.0	4.9	4.8	4.8	4.7	4.6	4.4
6	5.6	5.1	4.8	4.5	4.4	4.3	4.2	4.2	4.1	4.1	3.9	3.7
7	5.6	4.7	4.4	4.1	4.0	3.9	3.8	3.7	3.7	3.6	3.4	3.3
8	5.3	4.5	4.1	3.8	3.7	3.6	3.5	3.4	3.4	3.4	3.2	2.9
9	5.1	4.3	3.9	3.6	3.5	3.4	3.3	3.2	3.2	3.1	2.9	2.8
10	5.0	4.1	3.7	3.5	3.3	3.2	3.1	3.1	3.0	3.0	2.8	2.5
20	4.4	3.5	3.1	2.9	2.7	2.6	2.5	2.5	2.4	2.4	2.1	1.8
∞	3.8	3.0	2.6	2.4	2.2	2.1	2.0	1.9	1.9	1.8	1.6	1.0

Figure 16.3: Result of f-test to determine equivalence of variance for two experimental data samples (calculated in Excel).

Statistical software packages provide convenient *f*-test functions for determining equality of variance and provide a p-value (the chance that the variances of the two samples are equal).

A small p-value (<0.05) indicates that the chance that the variances of the two samples are equal is small, and that the variances are thus different.

For our sample data, a p-value of 0.25 (>0.05) thus confirms our table-based finding that the variances of the two samples are equal, and we may proceed to compare the two means using the two-sample *t*-test for equal variances. Alternatively, we would choose a modification of this test - the two-sample *t*-test for *unequal* variances.

Implementation of the *t*-test to determine if two samples (or a sample and a theoretical mean) have the same mean, or if there is a significant difference between the means, will be discussed in the next chapter.

---

1. The f-test was initially developed by Fisher as the variance ratio (defined as the ratio of explained variance (or between-group variability) to unexplained variance (or within-group variability) in the 1920s. The f-distribution was subsequently tabulated, and the f-test named, by Snedecor in 1934 as an improved presentation of Fisher's Analysis of Variance, in order to facilitate its interpretation within the biological sciences.

17

TWO SAMPLE MEANS COMPARISONS: THE T-TEST (PART II)

The test statistic for all *t*-tests (one sample, paired two-sample or independent two-sample *t*-tests) is the *t*-value or *t*-ratio, which is the difference between the sample means (numerator) divided by the standard error of the difference (denominator)[1].

Expressed in other terms - the numerator is the effect size (signal) and the denominator is the variability (noise) while the *t*-value may be considered to represent the signal:noise ratio.

The *t*-values are calculated differently for each test:

For a one-sample *t*-test, the expected or theoretical mean is subtracted from the mean of the sample, and this is divided by the standard error of the sample. The degrees of freedom (df) used in this test are n–1. For a sample of 10 seedlings, the degrees of freedom are thus [10-1=9].

For paired two-sample *t*-tests, the difference between the sample means is divided by the standard error of the mean differences. As the difference between the sample means is the same as a single sample (the mean differences), the degrees of freedom (df) used in this test are n-1. For two paired samples of 10 seedlings, the degrees of freedom are thus [10-1=9].

For two-sample *t*-tests, the difference between the sample means is divided by the pooled (averaged) standard error for the two samples (which in turn explains the need for an assumption of equal variances for the two samples). The degrees of freedom (df) used in this test are [n (sample 1) + n (sample 2) –

2]. For two samples of 10 seedlings, the degrees of freedom are thus [10+10-2=8].

As statistical software packages (and most spreadsheets) provide convenient *t*-test functions, we will not delve deeper into the calculation of *t*, but rather focus on interpreting the results.

Example: Comparing treated and untreated seedlings.

In our sample data (see previous chapter) of the height of ten untreated seedlings compared to the height of ten herbicide-treated seedlings, the *f*-test was used to determine whether the variances of the two samples were equal or not.

As the *f*-ratio was smaller than the critical *f*-value, we could conclude that the variances of the two samples were equal and that we may compare the means using the *two-sample t-test for equal variances* and calculate or compute the *t*-value (Figure 17.1). Had the variances not been equal, we would choose the two-sample *t*-test for unequal variances.

Untreated	*Treated*
63	53
72	48
54	43
62	45
57	55
73	47
53	41
51	57
61	53
69	39

	Untreated	*Treated*
Mean	61.6	48.1
Variance	60.68	37.96
Observations	10	10
Pooled Variance	49.32	
Hypothesized Mean Difference	0	
df	18	
t Stat	4.29	
P(T<=t) one-tail	0.0002	
t Critical one-tail	1.734	
P(T<=t) two-tail	0.0004	
t Critical two-tail	2.10	

Figure 17.1: Result of two-sample t-test for equal variances (with p-values) to compare means for two experimental data samples (calculated in Excel; top)

To interpret the results, the calculated *t*-value can be compared to *two-tail critical t table values* (Figure 17.2) for n-2 degrees of freedom (the total sample size of 10+10-2=18df) to determine whether the sample means are equal (or significantly different).

df	p= .05 2-tailed	df	p= .05 2-tailed
1	12.71	13	2.16
2	4.30	14	2.14
3	3.18	15	2.13
4	2.78	16	2.12
5	2.57	17	2.11
6	2.45	18	2.10
7	2.36	19	2.09
8	2.31	20	2.09
9	2.26	30	2.04
10	2.23	50	2.01
11	2.20	100	1.98
12	2.18	∞	1.96

Figure 17.2: Critical t-values for p= 0.05; two-tailed.

As p=0.05 corresponds to a 5% (or less) chance of discarding the null hypothesis (the means are equal), *t*-values less than the critical table value mean the null hypothesis may not be discarded *i.e.* the samples have equal means.

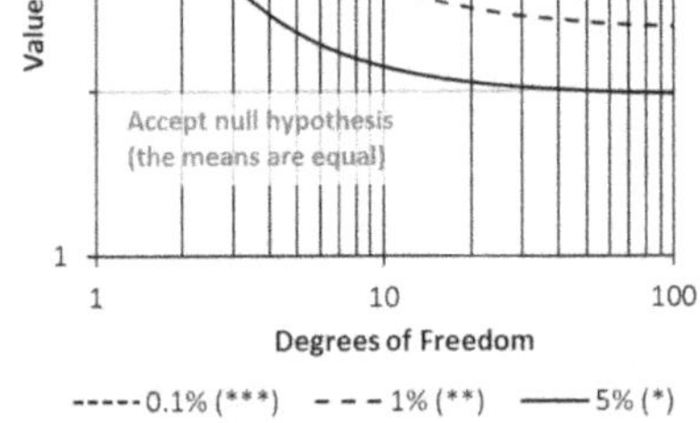

Figure 17.3: Values of t (two-tailed) for varying levels of significance.

In our example, the *t*-value (4.29) is greater than the critical *t*-value (*t*=2.10 for 18df), and we may thus conclude that the means of the two samples are significantly different.

Similarly, for our sample data a computed p-value of <0.05 (Figure 17.1) confirms that the means of the two samples are significantly different.

While *t*-tests allow us to determine if two samples have the same mean, biological experiments usually comprise more than two samples.

Accordingly, Analysis of Variance (ANOVA) tests are used to evaluate multiple (three or more) data samples, and this is the subject of the next chapter.

---

1. William Sealy Gosset (1876-1937), a chemist and statistician at the Guinness Brewery introduced the z-statistic to test the means of small, normally-distributed samples for quality control in brewing, and published it anonymously under the name "Student".
   Gosset's test was adapted by R.A. Fisher to incorporate the concept of "degrees of freedom" (introduced by him in its statistical context) in a 1924 paper, and Gosset appears to have introduced the t-nomenclature (t-distribution; t-test) in correspondence with Fisher. The t-statistic is also termed Students t-distribution and Students t-test.

18

MULTIPLE SAMPLE MEANS COMPARISONS: ANOVA TEST

While the t-test is used to compare the means of two treatments, samples or groups, Analysis of Variance (ANOVA[1]) are hypothesis tests used to differentiate between the means of more than two treatments, or groups.

Using ANOVA to compare two groups will return the same results (p-value) as using the t-test. Although it may seem simpler to compare the means of multiple samples as a series of multiple two-sample comparisons (t-tests), this is not a recommended option as increasing the number of comparisons compounds the error, and we risk making a type-I error *i.e.* falsely concluding a significant difference when there is no real difference.

The objective of the ANOVA test is thus to allow us to perform a single comparison test to compare multiple groups of data, thereby reducing the chance of making a type-I error.

As the mathematics required of the ANOVA tests are beyond the scope of this book, and as the tests can now be conveniently performed through spreadsheets, online calculators and dedicated statistical programs, we will in the following focus on understanding the underlying assumptions as well as the output of the ANOVA tests.

The most common application of ANOVA in crop protection R&D is the one-way ANOVA, used to differentiate between the means of more than two treatments, for which there is one measurement variable (*e.g.* plant height) and one nominal variable, or factor (*e.g.* pesticide).

<table>
<tr><th>TEST</th><th>TREATMENTS</th><th>FACTORS</th><th>EXAMPLE</th></tr>
<tr><td>t-test</td><td>Two treatments</td><td rowspan="2">One factor: e.g. pesticide treatment</td><td>Control vs. treated</td></tr>
<tr><td>One-way ANOVA</td><td rowspan="3">More than two treatments</td><td>Untreated control, treatment A; treatment B</td></tr>
<tr><td>Two-way ANOVA</td><td>Two factors: e.g. pesticide treatment and location</td><td>Three or more treatments, and two locations</td></tr>
<tr><td>Three-way ANOVA</td><td>Three factors: e.g. pesticide treatment, location and pest development stage</td><td>Three or more treatments, two locations with target pests from two development stages</td></tr>
</table>

Figure 18.1: Summary of Means comparisons.

Two- and three-way ANOVA tests are used to differentiate between the means of more than two treatments, for which there is one measurement variable and two (or three) nominal variables, or factors (Figure 18.1).

Assumption of Normality and Homogeneity of Variance

As is the case for the t-test, ANOVA assumes that the data is sampled from populations that have equal variance. Tests such as Bartlett's test for homogeneity of variance are typically used to test this assumption - if Bartlett's test reveals deviances from homogeneity, data transformation can be considered.

Both Bartlett's as well as the ANOVA test assume that the samples follow a normal (Gaussian) distribution (see previous chapters). A number of tests of normality are available, such as the Shapiro–Wilk and Kolmogorov–Smirnov tests, but these require statistical insight or access to statistical software packages.

If each treatment comprises less than ten data values, a general assumption (previously described as the Normality Rule of Thumb) is that any test of normal distribution will be so compromised that neither data transformation nor the use of nonparametric tests (*e.g.* the Kruskal-Wallis test, see chapter 12: "Nonparametric or Parametric Statistical Tests") will provide a significant benefit.

In daily practice, however, it is common for scientists to simply use parametric methods such as the ANOVA test directly on datasets from non-normal distributions.

Understanding the output of ANOVA tests

ANOVA differentiates between treatment (group) means by using the f-ratio to compare variance within and between treatments or groups - if the variation (typically termed "error") *within* groups is greater than the variation *between* groups, the difference between the means is not considered significant.

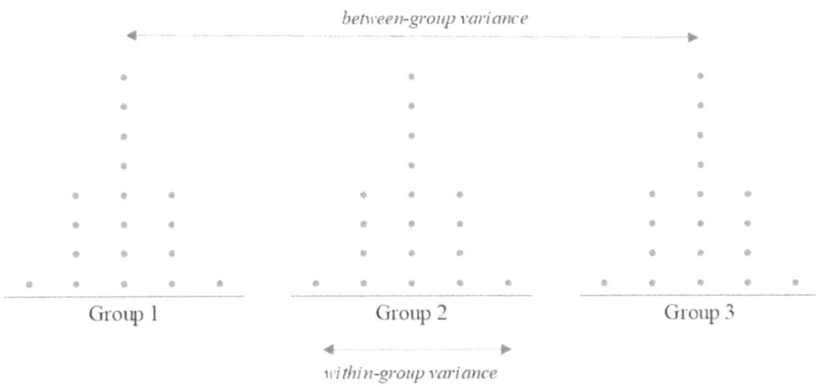

Figure 18.2: Between-group and within-group variance.

In the following example (Figure 18.3), we use the dataset for plant heights for ten untreated seedlings and ten herbicidally-treated seedlings (Treatment A) used in the previous chapter and expand it by adding an extra treatment (Treatment B), for a total of three groups (untreated control plus two treatments).

To calculate the within-group and between-group variance, the dispersion of the data (*Sum of Squares; SS*) was divided by the Degrees of Freedom (*df*) to give the variance (*Mean Square; MS*).

Degrees of freedom are calculated *between* groups (*m-1*) as 3 groups minus 1 = 2 and *within* groups (*n-m*) as 30 observations (*n*) minus 3 groups (*m*) = 27.

The f-ratio (F) is the ratio of the between-groups variance (*MS_between*) divided by the within-groups variance (*MS_within*), and the calculated ratio is compared to distribution values (one-tail critical f-table values) in order to determine significance (Figure 18.4).

Untreated	*Treatment A*	*Treatment B*
63	53	45
72	48	41
54	43	37
62	45	33
57	55	34
73	47	45
53	41	50
51	57	39
61	53	42
69	39	52

SUMMARY

Groups	*Count*	*Sum*	*Average*	*Variance*
Untreated	10	615.78	61.578	60.68
Treatment A	10	481.2	48.12	37.96
Treatment B	10	418	41.8	40.18

ANOVA

Source of Variation	*SS*	*df*	*MS*	*F*	*P-value*	*F crit*
Between Groups	2041	2	1020	22	0.0000021	3.35
Within Groups	1249	27	46			
Total	3290	29				

Figure 18.3: ANOVA output (Excel) comparing plant heights (cm) of untreated and treated plants.

As the f-ratio of the between-groups variance relative to the within-groups variance ($F = 22$) is greater than the table value of 3.35 (*F crt*) for 2 df between groups and 27 df within groups, we may conclude that there is a significant difference between the means of the treatments being compared.

For manual calculations of the f-ratio, significance may be determined using the f-table (figure 18.4).

However, statistical packages, spreadsheets and online ANOVA calculators supplement f-ratios and critical values with a P value (the chance that the means are equal).

If the p-value is less than 0.05, we may reject the null hypothesis that the means are equal and conclude that there is a significant difference between the means.

Likewise, if the p-value is greater than 0.05, we may accept the null hypothesis and conclude that there is no significant difference between the means.

df (within groups)	df (between groups or treatments)											
	1	2	3	4	5	6	7	8	9	10	20	∞
1	161	200	216	225	230	234	237	239	241	242	248	254
2	18.5	19.0	19.2	19.3	19.3	19.3	19.4	19.4	19.4	19.4	19.4	19.5
3	10.1	9.6	9.3	9.1	9.0	8.9	8.9	8.8	8.8	8.8	8.7	8.5
4	7.7	6.9	6.6	6.4	6.3	6.2	6.1	6.0	6.0	6.0	5.8	5.6
5	6.6	5.8	5.4	5.2	5.1	5.0	4.9	4.8	4.8	4.7	4.6	4.4
6	5.6	5.1	4.8	4.5	4.4	4.3	4.2	4.2	4.1	4.1	3.9	3.7
7	5.6	4.7	4.4	4.1	4.0	3.9	3.8	3.7	3.7	3.6	3.4	3.3
8	5.3	4.5	4.1	3.8	3.7	3.6	3.5	3.4	3.4	3.4	3.2	2.9
9	5.1	4.3	3.9	3.6	3.5	3.4	3.3	3.2	3.2	3.1	2.9	2.8
10	5.0	4.1	3.7	3.5	3.3	3.2	3.1	3.1	3.0	3.0	2.8	2.5
20	4.4	3.5	3.1	2.9	2.7	2.6	2.5	2.5	2.4	2.4	2.1	1.8
∞	3.8	3.0	2.6	2.4	2.2	2.1	2.0	1.9	1.9	1.8	1.6	1.0

Figure 18.4: One-tailed f-table for $\alpha = 0.05$

In our example above, we can determine that the p-value is considerably less than 0.001 (*i.e.* the difference between the means is extremely significant). However, the results of the ANOVA test only indicate if there is a significant difference between means, and do not reveal *which* group is different nor *how* it is different.

To determine this, we need to supplement the ANOVA test with post-hoc ("done after the event") or post-ANOVA tests for significant differences.

Post-ANOVA (post-hoc) tests

Post-ANOVA tests are typically performed after the f-ratio has indicated that not all means are equal.

Depending on our requirements, different post-ANOVA tests allow different comparisons, for example: to compare every mean to every other mean (the most common application), to compare each mean to a control mean, or to compare selected means.

Among the post-ANOVA tests to compare every mean to every other mean, we are provided with a number of methods, the most common of which (at least in the biological sciences) are the Tukey's test and the Newman-Keuls test (or Student-Newman-Keuls test; SNK).

The Newman-Keuls post-ANOVA test is considered less stringent (has less power - see previous chapter on Power) than Tukey's test, and may indicate that a difference in means is statistically significant, whereas the more stringent Tukey's test might indicate that the difference is not statistically significant.

As the risk of a Type-I error (where there really is no difference) is greater for the Newman-Keuls test than for the Tukey test, statisticians tend to recommend Tukey's test.

However, for Tukey's test the chance of a Type-II error (missing a real difference) is greater. This is the "one that got away" type error, also termed "false negative", where you miss out on something that really is extraordinary.

POST-ANOVA TEST	STRINGENCY	RISK
Tukey's test	More stringent: "there is no difference"	Type-II error (missing a real difference; false negative)
Newman-Keuls, or SNK	Less stringent: "there is a difference"	Type-I error (where there really is no difference; false positive)

Figure 18.5: Summary of post-ANOVA tests most commonly used to identify differences between all group means.

Taking the strategic research objectives into consideration will facilitate a reasoned choice of test method. Thus, if the cost of a false positive is high (for example if it leads to the initiation of a costly development process) a more stringent method to determine the significance of differences between treatment means such as Tukey's test may be appropriate.

Conversely, if the cost of a false negative is high (leading to you missing out on a valuable discovery) a less stringent method such as Newman-Keuls test may be appropriate.

Newman-Keuls test may thus be appropriate for early stage discovery processes, while Tukey's may be more appropriate as a decision-making tool as discovery candidates approach development phases.

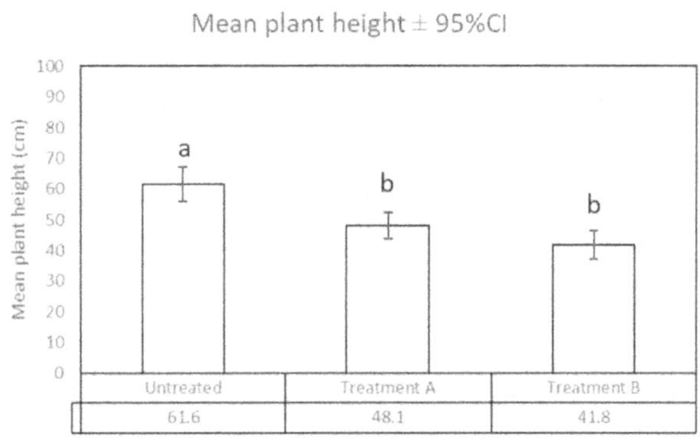

Figure 18.6: Mean plant heights for untreated control and two herbicide treatments. Different letters above bars indicate significant differences (p<0.05; Tukey's post-ANOVA test).

The means and results of post-ANOVA tests are typically presented as bar graphs (with error bars correctly defined), where different letters above bars (means) indicate significant differences ($p<0.05$). Conversely, bars labeled with the same letter are not significantly different from each other.

In contrast, applying the SNK test to the example dataset (Figure 18.7) confirms that the means of treatment A and Treatment B are significantly different from that of the untreated control ($p<0.001$) while the lower stringency of this tests indicates that the difference between the means of Treatment A and Treatment B is significantly different ($p<0.05$):

A number of additional post-ANOVA tests are available for specific comparisons (to compare each mean to a control mean, or to compare selected means, *e.g.* Bonferroni's test), or for situations for which the assumption of equal variances may (*e.g.* Duncan's test - a more stringent modification of the Newman-Keuls test *i.e.* with more power) or may not (*e.g.* Dunnett's test) be required.

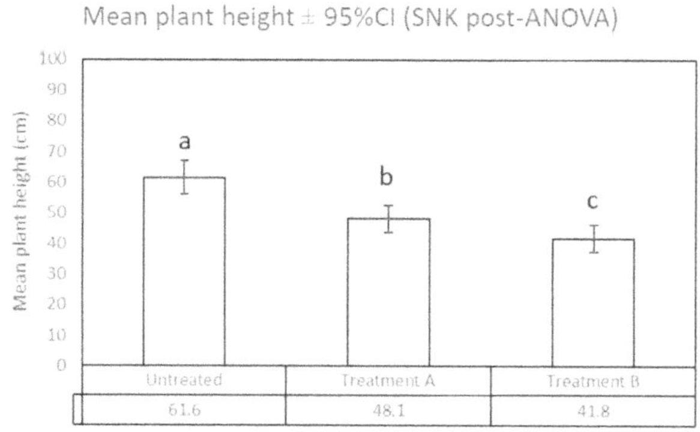

Figure 18.7: Mean plant heights for untreated control and two herbicide treatments. Different letters above bars indicate significant differences ($p<0.05$; SNK post-ANOVA test).

These necessitate an in-depth understanding of statistical analysis and are thus outside the scope of this book - the reader is encouraged to refer to a statistician.

---

1. One of British statistician and geneticist R.A. Fisher's (1892-1962) most important contributions to statistical methods is the analysis of variance (ANOVA), providing solutions to issues of variance arising from factors such as the heterogeneity of soils and the variability of biological material. Several of his most important contributions were published in "Statistical methods for research workers" (1925).

V

NONLINEAR AND LINEAR REGRESSION

19

NON-LINEAR REGRESSION (SIGMOIDAL DOSE-RESPONSES)

Bioassays (analytical methods to determine the potency of a substance by its effect on living organisms) are used in all fields of the biological sciences - from the agricultural sciences over toxicology and biochemistry to medicine - for the assessment of biological activity and selectivity of pesticides, biochemicals and pharmaceuticals.

Bioassays include dose-response studies using several (typically replicated) doses of one or more biologically active compounds, or active ingredient (a.i.) to measure the response of specific target organisms.

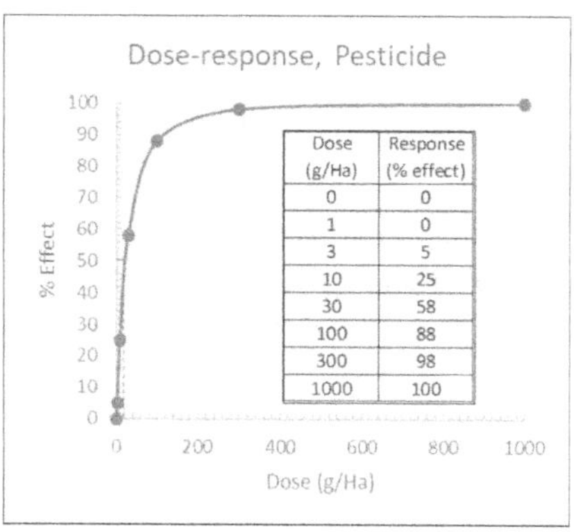

Figure 19.1: Hyperbolic dose-response (effect) curve for linear, arithmetic doses. Insert: Dose-response data table.

In dose-response experiments, an active ingredient's (*e.g.* herbicide) effect on a target organism (plant) is typically investigated by constructing a dose-response curve which describes the relationship between increasing the dose and the associated change in response across a dose range.

On an arithmetic (linear) scale, the dose is plotted on the X-axis and response is plotted on the Y-axis, and the response curve is hyperbolic.

On this linear scale, it is apparent that evaluating the transition from low to high response is difficult, as most of the information is compressed in a small section of the graph.

To address this, the doses on the X-axis may be logarithmically transformed.

On a logarithmic dose scale (see Figure 19.2), dose-responses typically follow the classic sigmoidal curve familiar to biologists from *e.g.* bacterial growth studies.

Sigmoidal dose-response fitting

From the approximately linear slope of the sigmoidal dose-response fit (from approximately 10-20% effect to 80-90% effect, over about two log units) we can derive Effective Dose levels (ED*x*), *i.e.* the doses causing a specific response *x* in the target organism.

Among these are the Effective Dose, ED50 (for example, herbicide dose required to reduce for example plant growth by 50%, *i.e.* half-way between the upper and lower asymptote; see Figure 19.2), ED10 (dose required to reduce plant growth by 50%, often defined as the No Observable Effect Level; NOEL) and ED90.

Other terminology used includes EC (Effective Concentration), LC/LD (lethal concentration/dose, for toxicology) and IC (inhibitory concentration, for biochemistry).

Pesticide dose-response experiments typically use 5-7 doses, equally spaced on a logarithmic scale. For example, doses might be 1, 3, 10, 30, 100, 300, 1000 g ai/Ha. When converted to logarithms, these values are equally spaced: 0.0, 0.5, 1.0, 1.5, 2.0, 2.5, 3.0.

This provides a sufficiently broad and equally spaced range of doses to ensure zero percent effect at the lower asymptote (plateau) and 100% effect at the upper asymptote - requirements for fitting the sigmoidal dose-response curve (see below).

For some trials (especially insecticide trials) mortality may arise for the untreated "control" due to natural causes, and Schneider-Orelli's formula may be used to correct for control response (typically mortality) if control response is more than 10% (see also chapter 12)

In its basic form, the standard sigmoidal dose-response curve showing response (Y) as a function of the logarithm of concentration (X) may be fitted by eye or by the following equation:

$$Y=Bottom+\frac{(Top-Bottom)}{(1+10^{\wedge}(logEC50-x)}$$

...where "Top" is the response at the upper asymptote (the response at high doses), "Bottom" is the response at the lower asymptote (response to the untreated control) and logEC50 is the logarithm of the EC50 (the dose giving a response half-way between the upper and lower asymptotes).

The number 1 indicates the standard slope (termed Hill slope) where the response increases linearly from approximately 10-20% to 80-90% of maximal as the dose increases over about two log units.

Dose (g/Ha)	log Dose	Response (% effect)
0.1	-1	0
1	0	0
3	0.5	5
10	1	25
30	1.5	58
100	2	88
300	2.5	98
1000	3	100

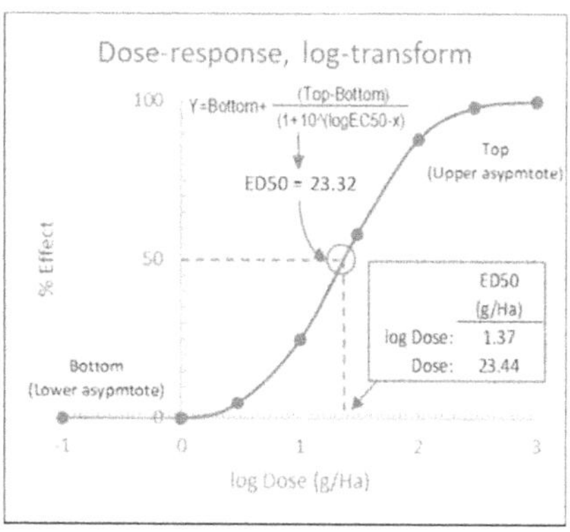

Figure 19.2: Sigmoidal dose-response (effect) curve for log-transformed doses, visual estimation and calculation of ED50. Top: transformed dose-response data table.

In our example above, using values from a herbicide dose-response curve, the

ED50 value may be fitted by eye as logED50 = 1.37 from which the dose may be determined as ED50 = 10^1.37 = 23.44g ai / Ha.

Alternatively, the ED50 value may be calculated (using *e.g.* a statistics package) as ED50 = 23.22g ai / Ha.

Here, both values are almost equal, as the data conveniently fits the sigmoidal curve - for data with more scatter around the regression line, visual determination of ED50 becomes less precise.

A further advantage of using statistical packages is that goodness-of-fit of the data to the regression curve can be quantified.

The standard sigmoidal dose-response curve assumes a standard Hill slope, but in practice, a revised equation for variable slopes is often used.

Theoretically, the slope should be equal when evaluating different formulations of the same active ingredient, and different when evaluating different formulations of the same active ingredient.

Often, experimenters will not have enough data to allow accurate curve fitting, and confidence intervals of the fit may be used to approach best-fit slope parameters. Alternatively, an extrapolated data value (typically 2 or three orders of magnitude higher than the expected 100% efficacy rate may be extrapolated.

These advanced cases (including biphasic dose-response curves) lie outside the scope of this book, but most statistical packages are able to fit these conditions.

If statistical software packages are not available, and experimenters do not have a background in mathematics, linear dose-response fitting may be conveniently performed using PROBIT models. This will be the topic of the next chapter.

20

LINEAR REGRESSION (PROBIT)

Probit ("probability unit") models were developed by American biologist and statistician Chester Bliss in 1934, as a method to evaluate dose response in pesticide data.

Probit allows researchers to convert mortality (effect) percentages to probit values, which approximate a straight line function between the logarithm of the dose and effect, and which can be analyzed by simple linear regression methods.

Probit is thus the transformation of the sigmoid dose-response curve to a straight line.

The Probit model was further adapted and tabulated at Rothamsted by British statisticians D. J. Finney and W. L. Stevens in 1948 to avoid having to work with negative probits in an era before the ready availability of electronic computing.

It is these Probit tables that even today ensure that dose-response fitting to evaluate dose-response relationships may be conveniently performed when statistical software packages are not available, and experimenters do not have a background in mathematics.

Probit analysis may be conducted using tables to determine the probits and fitting the relationship by eye or through linear regression, or by using a statistical package. The process for evaluating dose-response relationships through the Probit analysis by hand, or by using a spreadsheet, are outlined in the following:

Step 1: log transform the doses.

Step 2: Convert % response to probits (short for probability unit).

Probits are generally calculated in the range where the sigmoidal response increases linearly *i.e.* from approximately 10-20% to 80-90% of maxima and should ideally contain three points within this linear phase.

If control (untreated) response is more than 10%, Schneider-Orelli's correction (see previous chapter) may be used. Probits for a given percentage effect may be determined using Finney's table:

%		1	2	3	4	5	6	7	8	9
0	-	2.67	2.95	3.12	3.25	3.36	3.45	3.52	3.59	3.66
10	3.72	3.77	3.83	3.87	3.92	3.96	4.01	4.05	4.08	4.12
20	4.16	4.19	4.23	4.26	4.29	4.33	4.36	4.39	4.42	4.45
30	4.48	4.50	4.53	4.56	4.59	4.61	4.64	4.67	4.69	4.72
40	4.75	4.77	4.80	4.82	4.85	4.87	4.90	4.92	4.95	4.97
50	5.00	5.03	5.05	5.08	5.10	5.13	5.15	5.18	5.20	5.23
60	5.25	5.28	5.31	5.33	5.36	5.39	5.41	5.44	5.47	5.50
70	5.52	5.55	5.58	5.61	5.64	5.67	5.71	5.74	5.77	5.81
80	5.84	5.88	5.92	5.95	5.99	6.04	6.08	6.13	6.17	6.23
90	6.28	6.34	6.41	6.48	6.55	6.64	6.75	6.88	7.05	7.33

Figure 20.1 Finney's table for the transformation of response percentages to probits.

In our example, for a 25% response the corresponding probit is 4.33, for 58% effect probit=5.20 and for 88% effect probit=6.18:

Dose (g/Ha)	log Dose	Response (% effect)	Probit
0.1	-1	0	
1	0	0	
3	0.48	5	
10	1	25	4.33
30	1.48	58	5.20
100	2	88	6.18
300	2.48	98	
1000	3	100	

Figure 20.2: Dose-response data table, log-transformed doses and transformation of response percentages to probits.

Step 3: Graph the probits versus the log of the concentrations and perform a linear regression, by hand or using a spreadsheet (Figure 20.3).

Step 4: Determine the ED50.

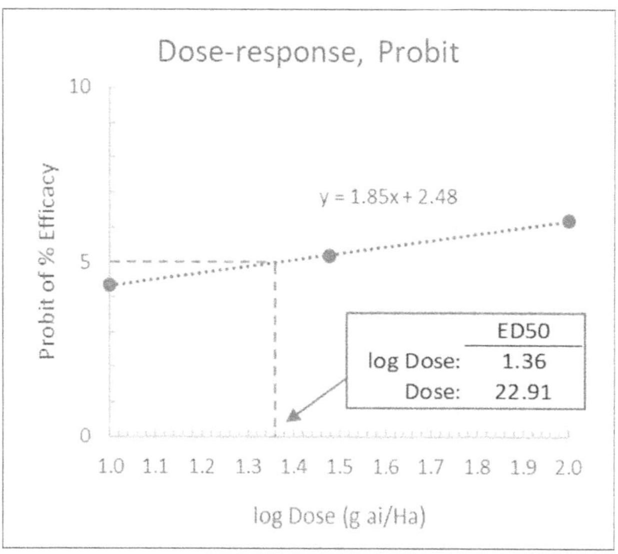

Figure 20.3: Linear fit of dose-response data, probits versus the log of the concentrations: visual estimation of ED50.

In our example, using values from a herbicide dose-response curve, the ED50 value may be fitted by eye as logED50= 1.36 from which the dose may be determined as ED50= 10^1.36= 22.91g ai/Ha:

Alternatively, if the linear function has been calculated (in our example as y= 1.85x + 2.48) the ED50 may be calculated as:

$$\text{log Dose} = \frac{\text{Probit value - intercept}}{\text{Slope}}$$

...where 2.48 is the intercept (at x=0) and 1.85 is the slope. The Probit value is 5 for ED50 (in Finney's table, Figure 20.1, for a 50% response (ED50) the corresponding probit is 5):

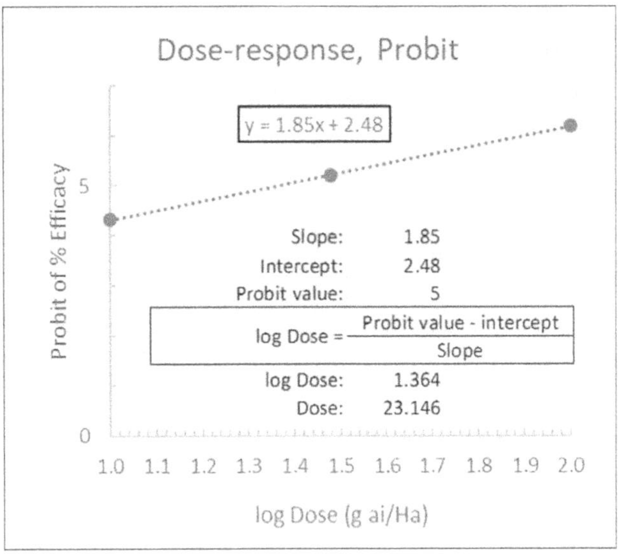

Figure 20.4: Linear fit of dose-response data, probits versus the log of the concentrations: calculation of ED50.

In our example, logED50= 1.364 from which the dose may be determined as ED50= 10^1.364= 23.146g ai/Ha.

An alternative method of calculating Effective Dose levels (ED*x*) is Logit, which can be performed through a similar process to that described for Probit.

The key difference between logit and probit models lies in the assumption of the distribution of the errors, where for probit the errors are assumed to follow a Normal distribution. In practice, both generally lead to the same conclusions and both are thus considered appropriate.

In our example, for both non-linear (sigmoidal) regression (see previous chapter) and linear (Probit) regression, both ED50 values are almost equal, and visual estimates were relatively accurate. For data with more scatter around the regression line, visual determination of ED50 becomes less precise.

As for non-linear regression, a further advantage of using statistical packages is that the goodness-of-fit of the data to the regression curve can be quantified.

21

AREA UNDER THE DISEASE PROGRESS CURVE (AUDPC)

Most plant pathogens are polycyclic, with recurring production of inoculum and reinfection leading to increases in the disease during the growing season.

By evaluating disease levels over time, it is possible to combine multiple observations from disease progress experiments into a single cumulative value (a measure of duration and intensity) by integrating the Area Under the Disease Progress Curve (AUDPC).

AUDPC is used in crop protection efficacy studies to compare levels of efficacy between treatments over time.

To obtain the AUDPC, disease evaluations commence when the symptoms appear, typically when the untreated control plots reach 20% disease severity and are repeated at 3-10 day intervals (days after treatment, DAT) until disease severity passes 80%, to ensure data is available for low, medium and high disease severity.

Ideally the AUDPC would be calculated through integration, although the midpoint method provides a simple estimate of the area under the infection curve.

The midpoint method approximates the area between the infection curve and the x-axis by summing the areas of rectangles with midpoints that are points on the curve between evaluations times.:

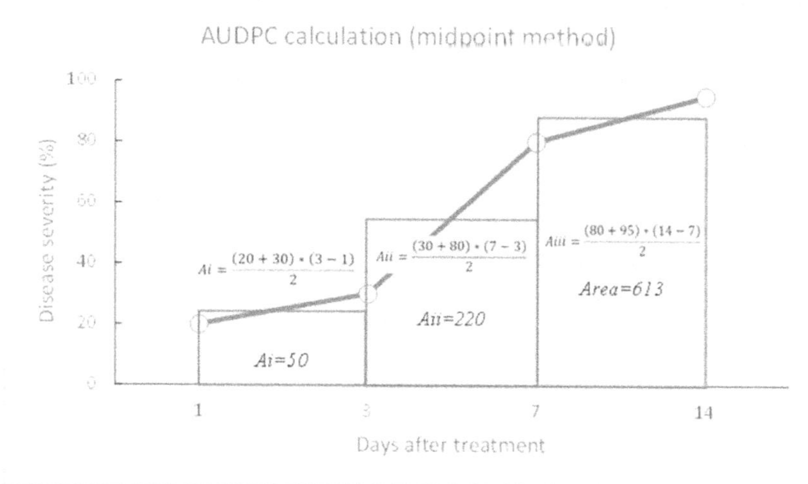

Figure 21.1: the midpoint method provides a simple estimate of the area under the infection curve.

The midpoint method is expressed as:

$$A(UDPC) = \frac{(SEV1 + SEV2) * (T1 - T2)}{2}$$

where SEV = disease severity and T = evaluation times (DAT) for two evaluations typically made at 3-10 day intervals.

<u>Example: calculating AUDPC using the midpoint rule</u>

For a treatment A, the following disease severity evaluations were made at four timepoints following application of treatment A:

	Days after treatment (DAT)			
Treatment	**1**	**3**	**7**	**14**
A	20	30	80	95

Step 1) add the first and second score (20 and 30)

Step 2) multiply by the time between them (50 x 2 days=100) and divide by two (=50).

Step 3) Repeat for second and third score: 30 + 80 = 110. 110 x 4 = 440.

Step 4) Divide by 2 = 220.

Step 5) Repeat for all consecutive pairs of scores and add the values together to determine the AUDPC:

$$Ai = \frac{(20 + 30) * (3 - 1)}{2} = 50$$

$$Aii = \frac{(30 + 80) * (7 - 3)}{2} = 220$$

$$Aiii = \frac{(80 + 95) * (14 - 7)}{2} = 613$$

$$AUDPC = Ai + Aii + Aiii = 883$$

Example: comparing treatments over time using AUDPC

In this example, the efficacy of three fungicides (ABC) is compared at multiple timepoints over the duration of the infection by evaluating disease severity at 1, 3, 7 and 14 days after treatment:

	Days after treatment (DAT)			
Treatment	**1**	**3**	**7**	**14**
A	25	25	40	97
B	20	20	70	97
C	30	45	71	97

The data can be graphically presented as:

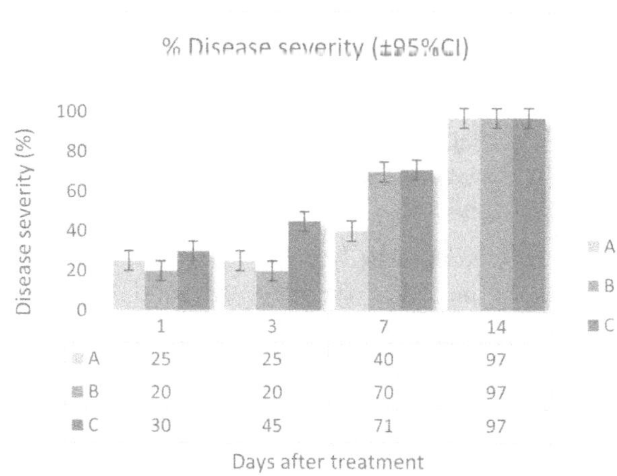

Figure 21.2: Results of efficacy trial, multiple observations from disease progress experiments.

For trials with multiple treatments and evaluations, it can be beneficial to create an overview and rank the treatments by combining multiple observations from disease progress experiments into a single, cumulative AUDPC value:

	Days after treatment (DAT)				AUDPC			
Treatment	**1**	**3**	**7**	**14**	1-3DAT	3-7DAT	7-14DAT	**Total**
A	25	25	40	97	50	130	480	660
B	20	20	70	97	40	180	585	805
C	30	45	71	97	75	232	588	895

In this example, treatment A is the most effective (lowest AUDPC) while treatment C is the least effective (greatest AUDPC).

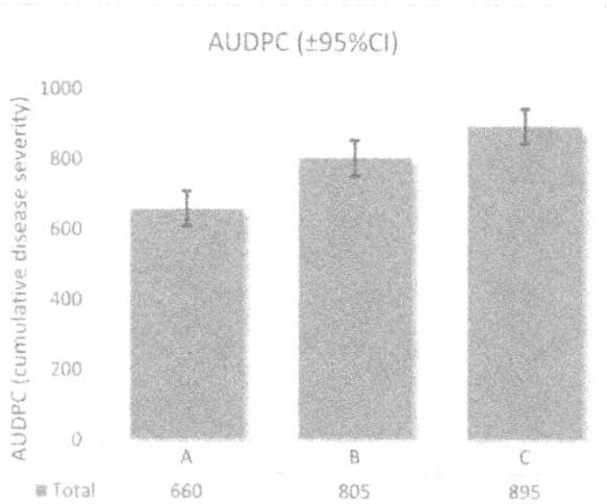

Figure 21.3: Results of efficacy trial, AUDPC.

While the advantage of AUDPC is to combine multiple observations of disease progress into a single value, this method tends to underestimate the effect of the first and last observations (upper and lower asymptotes).

This is a characteristic of the sigmoid curve and was identified in the previous chapter for EC50 calculation and will be reencountered in the following chapter on synergy.

This shortcoming may be alleviated by restricting the AUDPC calculation to the range between 20% and 80% disease severity i.e., to the linear phase of the curve.

VI

SYNERGY

22

DEMONSTRATING SYNERGY FROM EFFICACY DATA

Active ingredient mixes can extend the patent life of mixture partners and thus defend the market position. A synergistic effect is defined as whenever the action of a combination of active ingredients is greater than the sum of the actions of the individual components.

Not only is synergism an accepted argument against any allegation of obviousness in patent applications, but active ingredient synergism also allows better pest, pathogen or weed control with equal or reduced active ingredient rates. However, there is a risk that synergism could cause injury to crops or other non-target plants.

Synergy vs. formulation differences

Efficacy and synergy are determined not only by the active ingredients but also by the absence or presence of formulation inerts - solvents, adjuvants etc. Ideally, the individual active ingredients and the mix of actives should be formulated in the same way, to eliminate the effects of differing formulation inerts on biological activity.

In practice, however, it is common to observe mixes of formulated commercial products in patent applications, despite the issues of potential increased effect due to the combination of adjuvants in the commercial products - if there is no alternative, be prepared to demonstrate that differing formulation inerts have no effect on biological activity.

• • •

Models used to demonstrate synergy

In crop protection, two reference models are commonly used to demonstrate synergy: the Multiplicative Survival Model (MSM) for mixtures with differing action, and the Additive Dose Model (ADM) for mixtures with similar action.

- The **Multiplicative Survival, MSM** (or Independent Action, IA) models assume that the effects of active ingredients originate from different modes of action, where independent action implies there is no interaction between the active ingredients. These models propose that the individual effects of the ingredients should be multiplied together to estimate the overall effect of their combination. They tend to be more simple to perform, and are often preferred for patents and discovery trials. However, this model is subject to errors arising from the nonlinear nature of the dose-response curve, if the active ingredients are compared at rates outside the linear phase of the curve.

- The **Additive Dose, ADM** (or concentration addition, CA) models assume that the combined effects of active ingredients arise from similar modes of action (i.e. the active ingredients may target the same biological pathway or process but can have subtle differences in their specific molecular targets or modes of interaction (i.e. bind differently to the same target sites, and/or differ in potency). These models assume interaction between the active ingredients, and propose that the effects of individual ingredients can be added together to predict the overall effect of their combination. They tend to be more complex to perform, and are often preferred for development trials, where the ratio optimization is required.

Model	Assumptions	Analysis	Comment
Multiplicative Survival Model (MSM)	Assumes effects originate from different modes of action. Based on % mortality.	Abbot's formula (= Limpel's formula or Colby's method)	(+) Rapid and simple: does not require multiple doses / dose-response curves (and thus less experimental work) to predict expected responses. (-) Subject to errors arising from the nonlinear nature of the dose-response curve
Additive Dose Model (ADM)	Assumes effects originate from similar modes. Based on observed effective concentrations (e.g., EC50).	Wadley's method Tammes method Co-toxicity method	(-) Slower and more complex: requires multiple doses / dose response curves for EC50 determination (more experimental work) (+) Avoids errors arising from the nonlinear nature of the dose-response curve (+) optimization of mix ratios for maximal synergy

The choice between these models depends on the underlying assumption about the relationship between the modes of action of the active ingredients.

If the active ingredients are known to act in a similar manner, such as targeting the same biological process or pathway, the ADM models are more appropriate. Conversely, if the active ingredients are known to act through different mechanisms or pathways, the MSM models would be more suitable.

When evaluating mixes of active ingredients with similar modes of action, there may be instances where using the simpler Multiplicative Survival (MSM) models can be justified:

- During the initial screening or early stages of research, where the primary objective is to identify potential combinations for further investigation, the simpler models can serve as a starting point. These models allow for a quick assessment of the effects of the active ingredients and can guide decision-making on which combinations warrant more detailed examination.

- Using more complex models that account for interactions between active ingredients may require additional data, resources, or expertise. If these requirements are not readily available or feasible, adopting the MSM as a pragmatic approach can provide reasonable estimates of the combined effects.

However, it is important to note that while the MSM models offer simplicity and practicality, they may oversimplify complex biological interactions.

If there is evidence or strong indications of interaction or synergy between the active ingredients, employing more sophisticated models or conducting additional studies to explore these interactions would be necessary for a comprehensive assessment.

The synergy ratio (R)

To determine synergy, the synergy ratio (R) is calculated. If the experimentally observed efficacy of a mix of two active ingredients A and B ($A+B_o$) is greater than the expected, calculated expected efficacy ($A+B_e$) then synergy (R>1) is exhibited.

	Mathematical definition (MSM)	Biological definition (ADM)
Synergistic	R>1	R>1,5
Additive	R=1	R =0,5-1,5
Antagonistic	R<1	R<0,5

If R=1 then the effect is additive and if R<1 then antagonism is exhibited. For the more conservative MSM model (for patents and discovery trials, where the cost of a false negative is high), potential synergies are accounted for more cautiously, and the mathematical definition is used and accepted, while for ADM models (for development trials, where the cost of a false positive is high) a more stringent biological definition is generally used.

23

MULTIPLICATIVE SURVIVAL MODEL (MSM) FOR ACTIVE INGREDIENTS WITH DIFFERING MODES OF ACTION

In crop protection R&D, the Multiplicative Survival Model (MSM) for active ingredients with differing modes of action is most encountered (particularly in the patent literature), because they do not require multiple doses (and thus less experimental work) to predict expected responses.

Colby's method for binary mixes

In the following, we will refer to this method as Colby's method, as it is the most used name (despite Abbot's formula predating Limpel and Colby's later analysis of the method).

Colby's method is typically used in patent applications to provide a relatively simple demonstration of the interaction between active ingredients in a mixture (additivity, synergism, or antagonism).

For Colby's equation, calculations for synergy are best given from data where the requirement of dose-response linearity (generally considered to be within the range of 20% and 80% efficacy) is fulfilled - at higher rates the accuracy of synergy calculation is compromised due to the flattening of the dose-response curve at the upper asymptote. For Colby's equation, calculations for synergy are thus best given from data where the individual active ingredients, as well as the mixture, have been applied at a dose providing between 20-80% (ideally no more than 50%) control, since the requirement of dose-response linearity is fulfilled in this range.

In the Colby method, for a given combination of two active components, E (expected efficacy) can be expressed as:

$$E = A + B - (\frac{AB}{100})$$

where E = expected efficacy, A and B = the efficacy of two active ingredients A and B at a given dose.

Example, Colby's method for binary mixes:

100g/Ha of active ingredients A kills 30% of the weeds, 75g/Ha of B kills 50% of the weeds. If 100g of A kills 30% of the weeds, then there are 70% remaining. The best that adding 75g of B can expect to achieve is 40% of the remaining 70%. i.e., A+B(expected) =30+(40*70/100) = 58. If A+B (observed) is > A+B (expected) then synergy (R>1) is exhibited. If R=1 then the effect is additive and if R<1 then antagonism is exhibited:

100g/Ha	75g/Ha	100+75g/Ha		
A	**B**	**A+B (o)**	**A+B(e)**	**R**
30	40	80	58	1,38

In this example the observed efficacy is 80% for the mix of A+B, and the synergism ratio (R) between observed and expected is 1,38. The mix of A+B is synergistic. Coby's method can also be represented graphically as:

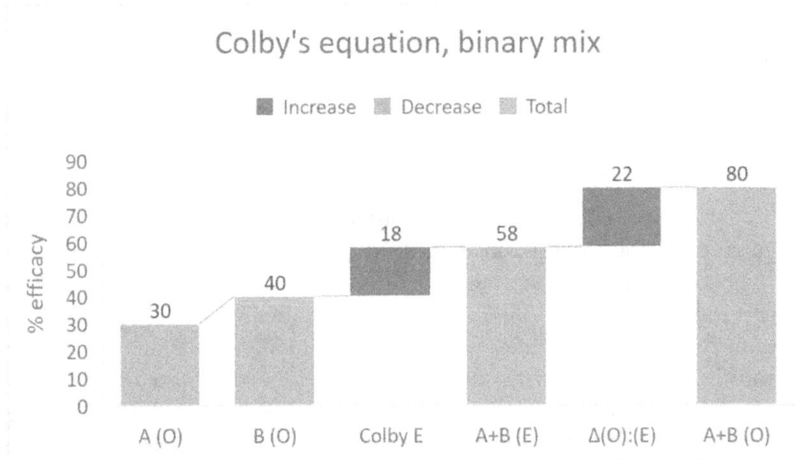

Figure 23.1: Colby's method for determining synergy, binary mixes.

1.2. Colby's method for ternary mixes

The MSM model for three active ingredients with differing modes of action is less commonly encountered. For a given combination of three active components, E (expected efficacy) can be calculated as:

$$E = A + B + C - \left(\frac{AB + AC + BC}{100}\right) + \left(\frac{ABC}{10000}\right)$$

where E = expected efficacy, A, B and C = the efficacy of three active ingredients A, B and C at a given dose.

When the percentage of control observed for the combination is greater than the expected percentage, there is a synergistic effect - the synergism ratio (R) is calculated as the ratio between the expected values and observed values.

If the synergism ratio (R) between observed and expected is >1 then synergy is exhibited, if R=1 then the effect is additive and if R<1 then the mix is antagonistic.

Example, Colby's method for ternary mixes:

1kg/Ha of active ingredients A kills 20% of the weeds, 1kg/Ha of B kills 30% of the weeds and 1kg/Ha of C kills 35% of the weeds. If A+B+C (observed) is > A+B+C (expected) then synergy (R>1) is exhibited. If R=1 then the effect is additive and if R<1 then antagonism is exhibited:

1kg	1kg	1kg	1+1+1=3kg	A+B+C (e)	R
A	B	C	A+B+C (o)		
20	30	35	80	64	1,26

In this example, the observed efficacy is 80% for the mix of A+B+C, and the synergism ratio (R) between observed and expected is 1,26. The mix of A+B+C is synergistic. The MSM method for ternary mixes with differing modes of action can also be represented graphically as:

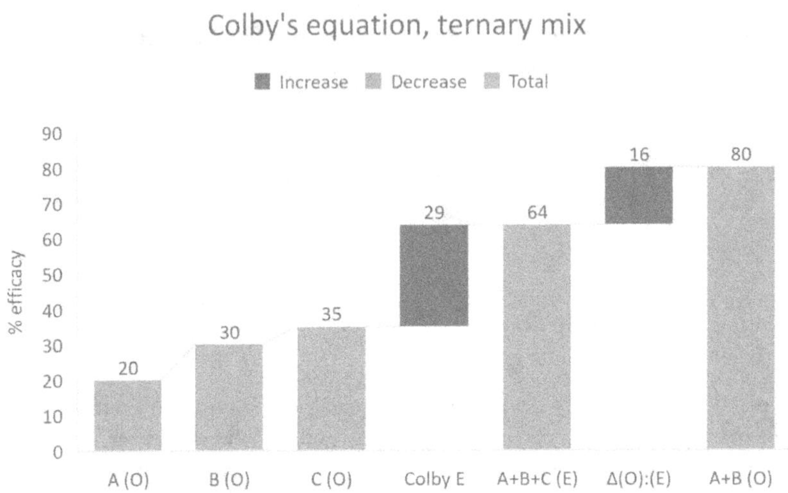

Figure 23.2: Colby's method for determining synergy, ternary mixes.

24

ADDITIVE DOSE MODEL (ADM) FOR ACTIVE INGREDIENTS WITH SIMILAR MODES OF ACTION

A more accurate (but less common, as it requires more experimental work) determination of the interaction between active ingredients in a mixture can be made by the Additive Dose Model (ADM).

The ADM method assumes that effects originate from similar modes of action, and that one component can substitute at a constant ratio for the other (the "equally effective rate").

The expected effectiveness of a mixture A+B(e) can thus be predicted from the effectiveness of the individual active ingredients when their relative ratio is known.

Example, equally effective rates:

In this example, the EC50 of active ingredient A is experimentally determined as 250g/Ha, the EC50 of active ingredient B is 0,9g/Ha.

In a mix, A and B substitute for each other at a constant ratio to give an "equally effective rate" – for example in mix 2 (below): a mix of A+B where 50% of A (125g ai/Ha) is replaced by 50% of B (0,45g ai/Ha).

The "weight ratio" is defined as the ratio of the individual active ingredient (A or B) to the mix (A+B). Examples of other ratios are given in the following, and will be the basis of the following method examples:

"Equally effective" rates, A+B (g ai/Ha)

Mix	AI	Ratio	g ai/Ha	Weight ratio
1	A	75%	188	99,9%
	B	25%	0,23	0,1%
2	A	50%	125	99,6%
	B	50%	0,45	0,4%
3	A	25%	63	98,9%
	B	75%	0,68	1,1%

Wadley's method for binary mixes

Wadley's method is based on EC50 values (effective concentration required to obtain a 50% effect) typically derived by linear or non-linear regression from dose-response experiments, and thus avoids errors arising from the nonlinear nature of the dose-response curve.

In the Wadley method, for a given combination of two active components, E (expected EC50) can be expressed as:

$$E = \frac{(a + b)}{\frac{a}{EC50(Ao)} + \frac{b}{EC50(Bo)}}$$

where a and b are the weight ratios (EC50 values in g/Ha) of compound A and B in the mixture and EC50(Ao) and EC50(Bo) are the experimentally observed (o) EC50 values obtained using the dose response curves for the individual compounds.

The synergism ratio (R) is calculated as the ratio between the expected values and observed values, EC50 A+B(e)/EC50 A+B(o).

Example, Wadley's method for binary mixes:

The EC50 of active ingredient A is experimentally determined as 250g/Ha, the EC50 of active ingredient B is 0,9g/Ha. In a mix, A and B substitute for each other at a constant ratio where 50% of A (125g ai/Ha) is replaced by 50% of B (0,45g ai/Ha).

The EC50 is 70 g ai/Ha for the mix of A+B. If A+B (observed) is > A+B (expected) then synergy (R>1) is exhibited. If R=1 then the effect is additive and if R<1 then antagonism is exhibited:

EC50	EC50	EC50	EC50	R	Weight ratio	
A(o)	B(o)	A+B(o)	A+B(e)		a	b
250	0,9	70	125,45	1,79	99,64	0,36

In this example, the observed EC50 is 70 g ai/Ha for the mix of A+B and the synergism factor (R) between observed and expected is 1,79. The mix of A+B is synergistic.

One objective of studying synergy is to maximize the cost/benefit ratio of active ingredients in a mix, to attain maximal efficacy with minimal active ingredient.

Wadley's method can be used to evaluate multiple "equally effective rate" ratios (see above):

EC50	EC50	EC50	EC50	R	Weight ratio	
A(o)	B(o)	A+B(o)	A+B(e)		a	b
250	0,90	110	187,73	1,71	99,9%	0,1%
250	0,90	70	125,45	1,79	99,6%	0,4%
250	0,90	40	63,18	1,58	98,9%	1,1%

In this example, the synergy ratio (R=1,79) is greatest for the 50:50 mix, with A and B substituting for each other at a constant ratio where 50% of A (125g ai/Ha) is replaced by 50% of B (0,45g ai/Ha).

Wadley's method for ternary mixes

For ternary (three component) mixes, Wadley's theoretical or expected model may be rewritten as:

$$E = \frac{(a+b+c)}{\frac{a}{EC50(Ao)} + \frac{b}{EC50(Bo)} + \frac{c}{EC50(Co)}}$$

where a, b and c are the weight ratios (ratio of EC50 values in g/Ha) of compounds A, B and C in the mixture and EC50(Ao), EC50(Bo) and EC50(Co) are the experimentally observed (o) EC50 values obtained using the dose response curves for the individual compounds.

The synergism ratio (R) is calculated as the ratio between the expected values and observed values, EC50 A+B+C(e)/EC50 A+B+C(o).

Isobole method for binary mixes and multiple dose ratios

The Isobole method (also called Tammes method) is an extension of the Wadley method and uses a graphic representation to determine the effect of multiple dose ratios on the level of synergistic effect, to determine the mix ratio at which the greatest synergy is observed.

Like the Wadley method, the Tammes method is based on EC50 values and avoids errors arising from the nonlinear nature of the dose-response curve.

Example, Isobole method for binary mixes with multiple dose ratios:

In this example, the EC50 of active ingredient A is experimentally determined as 250g/Ha, the EC50 of active ingredient B is 0,9g/Ha. In a mix, A and B substitute for each other at a constant ratio to give an "equally effective rate" – for example in mix 3 (below): a mix of A+B where 50% of A (125g ai/Ha) is replaced by 50% of B (0,45g ai/Ha).

The EC50s are used to plan a factorial trial with 25 treatments / 50 rates (g ai/Ha) comprising five rates (0,25X to 4X) and five mix ratios (100:0; 75:25; 50:50; 25:75; 0:100):

			"Equally effective" rate, A+B (g ai/Ha)				
Mix	**AI**	**Ratio**	**4x**	**2x**	**1x**	**0,5x**	**0,25x**
1	A	100%	1000	500	**250**	125	63
	B	0%	0	0	0	0	0
2	A	75%	750	375	188	94	47
	B	25%	0,90	0,45	0,23	0,11	0,06
3	A	50%	500	250	125	63	31
	B	50%	1,80	0,90	0,45	0,23	0,11
4	A	25%	250	125	63	31	16
	B	75%	2,70	1,35	0,68	0,34	0,17
5	A	0%	0	0	0	0	0
	B	100%	3,60	1,80	**0,90**	0,45	0,23

The factorial trial with 25 treatments is carried out and EC50 values are determined by e.g. nonlinear regression for each mix ratio, and the EC50s for each AI are calculated as a factor of their weight ratios (e.g. for mix 3 (above), the rates 250 : 0,9g ai/Ha corresponds to a ratio of 99,6 : 0,4%).

The outcome of mix 3 corresponds to the example given earlier for Wadley's method - the observed EC50 is 70 g ai/Ha for the "equally effective dose" of A+B (where 50% of A (125g ai/Ha) is replaced by 50% of B (0,45g ai/Ha).

Finally, the EC50s for each AI are plotted as an isobole (a plot in rectangular coordinates where the axes represent the doses of active ingredient A and B):

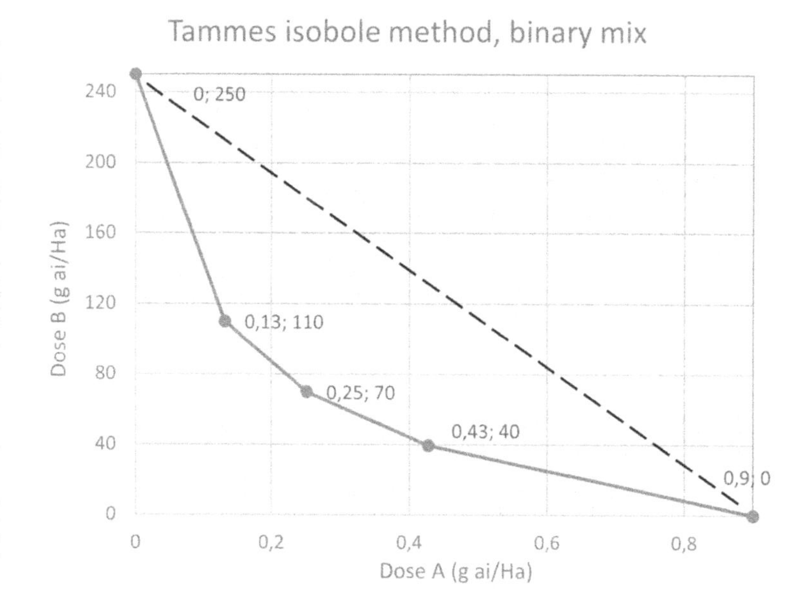

The dotted straight line represents the EC50 isobole for an expected additive effect (additive isobole) of differing "equally effective" ratios of A and B, while the curved full line represents the actual experimental EC50s for the specified mixture ratios.

If mixture EC50s are positioned below the additive isobole (as they are here) they are considered synergistic (synergistic isobole), and if the mixtures are positioned above, they are considered antagonistic (antagonistic isobole).

The mix ratio with the greatest distance from the isobole (in this example mix 3) has the greatest synergy. This corresponds to the Wadley analysis, where the synergy ratio (R=1,79) was greatest for the 50:50 mi, with A and B substituting for each other at a constant ratio where 50% of A (125g ai/Ha) is replaced by 50% of B (0,45g ai/Ha).

Co-toxicity model for binary mixes

A further method occasionally encountered in the literature is that of co-toxicity model. The outcome on the same dataset is the same as for the Wadley model (see below).

In contrast to the Wadley model, this model is based on converting EC50 values for the individual active ingredients and the mix to a toxicity index (TI):

$$TI(A)\ compared\ to\ B\ as\ the\ standard = \frac{EC50(A)}{EC50(B)} * 100$$

$$TI(B) compared\ to\ B\ as\ the\ standard = \frac{EC50(A)}{EC50(B)} * 100$$

$$Observed\ TI(A+B) = \frac{EC50(A)}{EC50(A+B)} * 100$$

$$Theoretical\ TI(A+B) = [TI(A) * \%\ of\ A\ in\ mix] + [TI(B) * \%\ of\ B\ in\ mix]$$

The cotoxicity coefficient (CC) of the mix can then be calculated as:

$$CC(A+B) = \frac{TI(A+B)\ observed}{TI(A+B)\ theoretical}$$

When the cotoxicity coefficient (CC) of the mix is greater than 100, there is a synergistic effect. If the CC equals 100 then the effect is additive and if less than 100, antagonism is exhibited).

Example, co-toxicity method for binary mixes:

The EC50 of active ingredient A is experimentally determined as 250g/Ha, the EC50 of active ingredient B is 0,9g/Ha and the EC50 is 70 g ai/Ha for the mix of A+B.

The theoretical toxicity index for the mix is 0,72 and the observed toxicity index for the mix is 1,3, giving a cotoxicity coefficient (CC) of the mix of 179:

EC50	EC50	EC50	Cotoxicity coefficient	R	Weight ratio		Toxicity index (TI)		Toxicity Index (A+B)	
A(o)	B(o)	A+B(o)	CC(A+B)		a	b	TI(A)	TI(B)	Observed	Theoretical
250	0,9	70	179	1,79	99,64	0,36	0,36	100	1,3	0,72

This outcome again corresponds to the example given earlier for Wadley's method - for both methods (Wadley and co-toxicity) the synergism factor (R) between observed and expected is 1,79 and the mix of A+B is synergistic.

STRATEGIC R&D MANAGEMENT

FOR PESTICIDE & BIOPESTICIDE R&D

PESTICIDE & BIOPESTICIDE INDUSTRY OVERVIEW

Global human population growth is around 75 million annually, and estimates place our total population at over 9 billion by mid-2050.

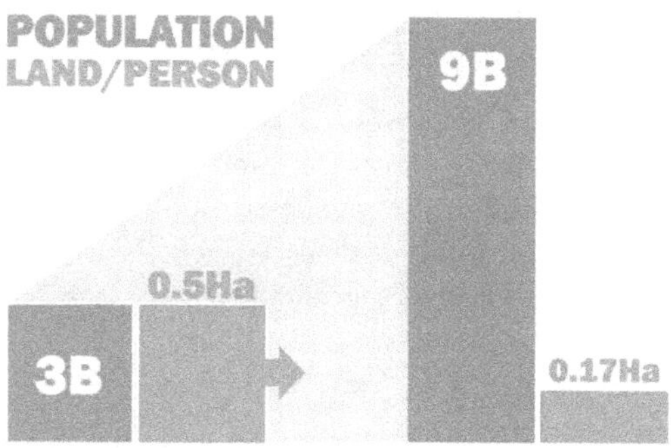

Figure I: Global population growth (1960 - 2030) vs. tillable land availability (Ha / person).

The logistical challenge is not only the projected 9 billion people, but also the estimated hundred billion head of livestock we will need to feed to maintain our current consumption standards.It has been estimated that there are only about 4 billion hectares of arable land in the world.

If 0.1 Ha of land is required to provide food for one person per year, then 4 billion Ha of tillable land would support 40 billion people.

With a 1% annual population growth rate, the carrying capacity of Earth may be reached in a little over two centuries.

Without crop protection strategies, we risk losing almost half of our overall crop production to insects, weeds and disease.

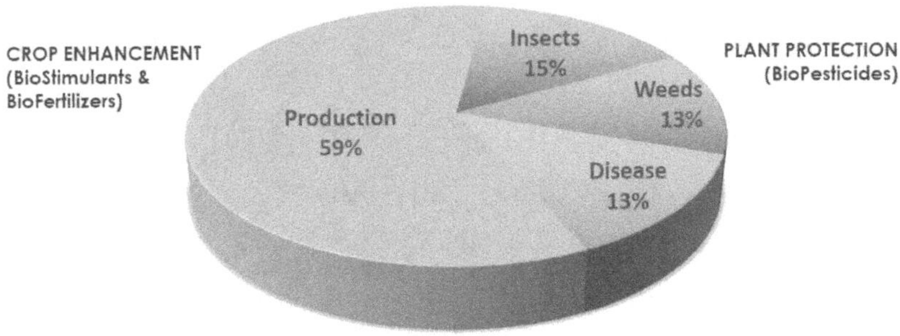

Figure II: Percentage of overall crop production at risk to loss by insects, weeds and disease.

Plant Protection agents - including Biopesticides - reduce these losses, while Crop Enhancement agents (including Biostimulants and Biofertilizers) contribute to increasing overall crop production.

To provide sufficient crop resources, new and innovative crop protection strategies are constantly being developed and implemented. These include formulation and application innovations for conventional pesticides (herbicides, insecticides and fungicides) as well as the development and application of economically viable biopesticides.

Pesticides

Pesticides are agents, substances or mixtures of substances used to prevent, destroy or control pests (including weeds) and diseases.

Pesticides may be classified by target (herbicides, insecticides, fungicides , etc.), grouped into chemical families (organophosphates, carbamates etc.) or classified according to their biological mode of action (photosynthesis inhibitors, respiration inhibitors etc.).

Herbicides – also known as weed killers – are pesticides toxic to plants that are used to destroy unwanted vegetation (weeds).

Herbicides control weeds by interfering with their growth, targeting specific enzymes and inhibit their normal function.

Insecticides are pesticides toxic to insects that are used to destroy unwanted pests, as well as their eggs and larvae. Insecticides control pests by interfering with their growth or reproduction, or by repelling insects from treated crops.

Fungicides are synthetic or natural chemical compounds or biological organisms able to kill or inhibit fungi or fungal spore germination, as well as oomycetes.

A pesticide may be a chemical (*conventional pesticide*) or a biological agent (*biopesticide*).

Within the EU, stricter pesticide regulatory requirements, consumer awareness regarding hazards caused by chemical pesticides and the phasing out of toxic conventional pesticides are factors driving the growth of the biopesticide market.

Biopesticides

Biopesticides, together with biostimulants and biofertilizers make up the product group of Agricultural Biologicals.

Biostimulants and *biofertilizers* are defined as biological substances or microorganisms capable of promoting crop health by increasing the availability of nutrients, or by acting as plant strengtheners and phytostimulants.

Biopesticides fall into three major classes:

Microbial pesticides, including the biopesticidal metabolites produced by these organisms

Bio-derived chemicals (metabolites) - naturally occurring compounds, including: *Semiochemicals* - natural compounds such as pheromones, ; *Biocontrol compounds isolated from microorganisms* (*e.g.* Spinosad); *Biocontrol compounds isolated from fungi or crustacean shells,* such as Chitosan; *Plant extracts,* including secondary biocontrol metabolites such as alkaloids, terpenoids and phenolics; and *naturally occurring minerals* (such as silicates) with biocontrol activity

Plant-incorporated protectants (PIPs) comprising foreign DNA inserted into crop genetic material (GM crops)

Biorational products (biorationals) are typically derived from natural or biological origins and include biological pesticides as well as products used for crop stress management, enhanced plant physiology benefits, and root growth management.

Bioherbicides include microorganisms that infect weeds, and natural compounds with herbicidal activity sourced from microorganisms or plants.

Bioinsecticides include microorganisms that infect insects, and natural compounds with insecticidal activity sourced from microorganisms, plants or animals.

Biofungicides are formulations of living organisms or natural metabolites used to control the activity of plant-pathogenic fungi.

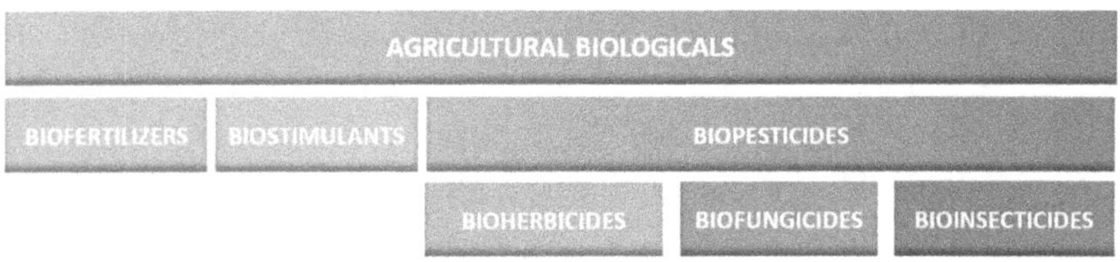

Figure III: Agricultural Biologicals.

By introducing unique modes of action, biopesticides can be used in integrated strategies with traditional pesticides to increase crop yields by working synergistically with chemical pesticides, to extend application timings and allow timely reuptake intervals as well as to provide resistance strategies.

Efficacy trials demonstrate conclusively that biology-based crop protection strategies can attain an efficiency at least on par with conventional crop protection strategies, when applied with due regard for factors such as plant-response kinetics and disease pressure development.

Most biopesticides are intended for use with traditional pesticides, which is contributing to the rapid increase in their use.

The leading AgChem companies are expanding their biopesticide R&D and portfolios, with focus on biochemicals, such as naturally occurring metabolites with entomopathogenic or antifungal activities, microbiome (beneficial microbes which improve nutrient uptake and disease resistance in crops), and molecular biological techniques (gene targeting) such as CRISPR and RNAi.

Integrated crop management (ICM) is a holistic approach to crop management, integrating agricultural and environmental management practices such as optimal site selection, crop rotation and optimization of crop nutrition, soil and water management as well as environment, energy and waste management and integrated crop protection strategies, in accordance with local conditions, climate and requirements.

Driven by legislation as well as consumer and industry demands, ICM strategies integrate local knowledge and experience with research and technological innovations to ensure sustainable agricultural production.

Integrated Pest Management (IPM) is a part of Integrated Crop Management (ICM) and the concept was introduced more than fifty years ago as a response to the preponderance of broad-spectrum, mainly neurotoxic conventional insecticides available to the market.

Currently, the definition of "Pest" in IPM includes vertebrate and invertebrate pests (e.g., birds, insects, mites and nematodes), plant pathogens (e.g., fungi, bacteria and viruses) as well as weeds.

The objective of IPM is to maintain pest populations below levels that would cause economic loss while minimally impacting the environment

IPM integrates several strategies including chemical, biological & mechanical control, precision technology and decision support tools (big data, including crowd-scouting and degree-day models for pest prediction) and the activation of natural plant defense responses as well as the use of resistant cultivars.

The aim is to reduce the use of chemical pesticides and manage the development of resistance in target pests to both chemical pesticides and biopesticide toxins.

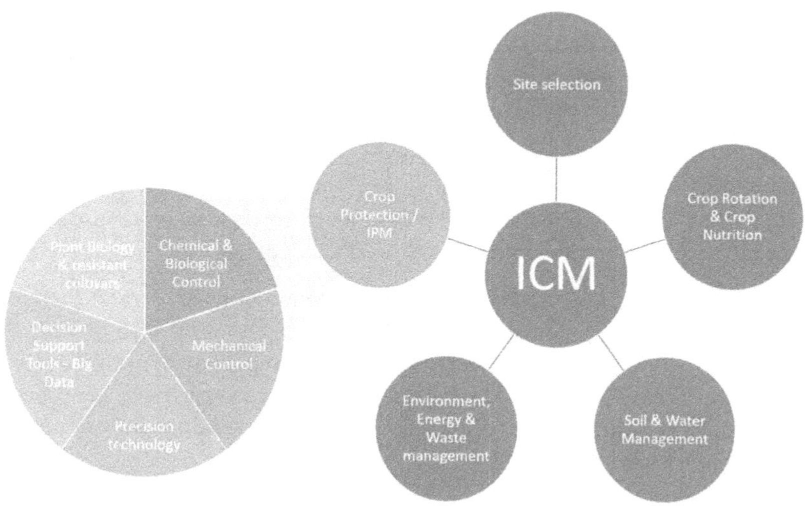

Figure IV: Integrated Pest Management (IPM) is a component of Integrated Crop Management (ICM).

By introducing unique modes of action, biopesticides are used in integrated strategies with traditional pesticides to increase crop yields by working synergistically with chemical pesticides, to extend application timings and allow timely reuptake intervals as well as to provide resistance strategies.

While the adoption of more 'biologically-based' IPM strategies in the field is still low - in part due to a perception of lower and variable efficacy, shorter persistence and greater dependence on environmental conditions on the part of end users - biological control plays an important role in the production of greenhouse crops. This is primarily due to their effectiveness under controlled environmental conditions, short persistence and low operator risk.

With advances in the development of biopesticides and technological developments in the use and application of conventional and biological pesticides, we now have a real opportunity to harmonize biological and chemical crop protection and realize the concept of IPM as a truly sustainable and holistic solution to current crop protection challenges.

While the post-war era of discovery and commercialization was a period of continuous innovation, since the 1980s crop protection corporations increasingly counted on growth through global expansion and reliance on revenues from existing products.

With the financial crisis of 2008, globalization stagnated, crop protection products became commodities, and corporations focused on "growth" through acquisitions, collaborations, and licensing agreements.

Many of these acquisitions and consolidations were driven by the need to spread the high fixed costs of pesticide R&D - including the discovery and development of new active ingredients - production, marketing and distribution over a greater sales volume in more markets (economy of scale).

Within the global crop protection industry, R&D resources have become too diluted for many companies to utilize true product innovation and strategic vision, and larger companies have leveraged consolidation to access the research and development initiatives of smaller, more innovative companies.

Recently however, R&D restructuring and investment by leaders in the crop protection industry has shown an increase. This is being driven by technological developments such as the introduction of novel active ingredients, the adoption of genetically modified crops, new application techniques, formulation developments and the introduction of unique biopesticidal product ranges.

I

THE CROP PROTECTION INDUSTRY

1

STRATEGIC TRENDS

Crop Protection Trends

Estimates of global market sizes for crop protection products (conventional pesticides as well as biopesticides) vary, but the consensus is that biopesticides are expected to increase from about 2% of the global pesticide market in 2003 to about 8% of the global pesticide market (estimated to exceed 82 billion USD) in 2020.

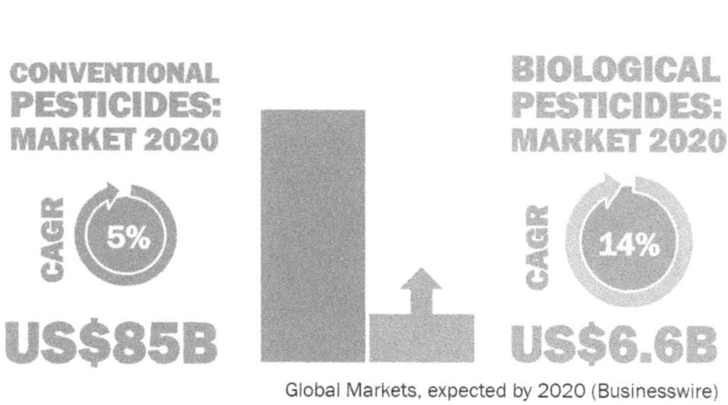

Figure 1.1: Market share and growth rates for Pesticides and BioPesticides.

Current opinion is that the compound annual growth rate (CAGR) for biopesticides during this period is expected to exceed 12%, compared to an expected overall pesticide CAGR of about 7%.

Falling commodity prices reduce the value of crops for farmers, and thus demand for crop protection products.

While lower oil prices permit cheaper production, associated declines in bioethanol margins restrict grower willingness to invest in these products.

Conventional Pesticides

Pesticide production is increasing to meet demand in rapidly expanding regions such as Latin America, Asia-Pacific and Africa - a potential overcapacity may, however, lead to a decline in prices.

FACTOR	TREND	COMMENT
SYNTHETICS		
Commodity prices	↓	↓ CPP demand
Oil	↓	Cheaper CPP but ↓ bioethanol margins
Regulation (EU)	↑	↓ CPP value
Regulation (BRA)	↑	
AIs (EU)	↓	↓ CPP potential
GMO	↑	↑ Dicamba; 2,4-D
Formulation dev.	↑	Seed treatments; stabilized Dicamba
Production	↑	↓ prices (overcapacity)
Demand	↑	China, India, Africa
NON-SYNTHETICS - "BIOLOGICALS"		
Biochemicals	↑	↑ increased R&D spend ↑ regulatory requirements – natural plant toxins?
Microbiome		
Gene targeting		

Figure 1.2: Strategic trends for Pesticides and Biopesticides.

A few active ingredients, such as Dicamba and 2,4-D are experiencing increased interest, due to trends in GMO technologies (trait stacking) and advances in formulation development (seed treatments).

Stricter pesticide regulatory requirements in the EU and Brazil restrict access and increase registration costs, while an associated decline in the number of registered active substances in the EU is reducing the potential of crop protection resistance strategies.

. . .

Biopesticides

Within the EU, stricter pesticide regulatory requirements, consumer awareness regarding hazards caused by chemical pesticides and the phasing out of toxic conventional pesticides are factors driving the growth of the biopesticide market.

The leading AgChem companies are expanding their biopesticide R&D and portfolios, with focus on biochemicals, such as naturally occurring metabolites with entomopathogenic or antifungal activities, microbiome (beneficial microbes which improve nutrient uptake and disease resistance in crops), and molecular biological techniques (gene targeting) such as CRISPR and RNAi.

Efficacy trials demonstrate conclusively that biology-based crop protection strategies can attain an efficiency at least on par with conventional crop protection strategies, when applied with due regard for factors such as plant-response kinetics and disease pressure development.

On the other hand, reports of inconsistent efficacy in the field lead to either over-optimism or skepticism by corporate decision makers and farmers alike.

From a grower / end-user point of view, an important strategic strength of biopesticides lies in their application at specific timings, such as pre-harvest application.

Short pre-harvest intervals allow the application of biopesticides (especially biofungicides) immediately before harvest, where few (if any) conventional pesticides are available.

Potential restraining factors include the current low adoption rates of biopesticides, as well as the expectation of stricter future regulatory requirements for biologicals.

Of importance with respect to toxicology, the formation of plant-produced toxins induced by biopesticides requires immediate attention.

The following is an overview of evolving trends and technologies which are giving biological pesticides an increasingly competitive edge:

Biologicals start-ups will continue to dominate innovation: Following recent corporate mergers, focus appears to be on integration, with less attention being placed on the acquisition of biological candidates or start-ups.

In contrast, biologicals start-ups are responding to increasing demands for biological pesticides, developing new solutions, and entering high-value markets. As corporations return their focus to innovation and R&D activities, these start-ups can be expected to be targets for acquisition.

Solo biologicals will increasingly be replaced by integrated solutions: While solo biological products have been the focus of development strategies, the need to improve the efficacy of biologicals is driving attention to integrated approaches, such as the inclusion of biologicals in IPM integrations, and the optimization of biologicals application through field-level monitoring and precision ag, leveraging the ability to monitor pest and pathogen dynamics in the field.

Together, conventional and biological combinations can provide the required level of pesticidal efficacy while allowing a reduction in the use of chemical actives as well as reducing the risk of resistance. However, registration challenges can be expected to curb the development of combination chemical and microbial products in favor of tank-mix solutions.

Recent advances in agrobiological knowledge will increase biologicals efficacy and reduce variability: Improving biologicals efficacy and reducing their variability will be facilitated by the increased implementation of agrobiological knowledge, through an improved understanding of biologicals mode-of-action as well as the integration of inducible plant defense responses.

The ability of growers to combine the use of biologicals with the ability of crops to induce defense responses will depend on an understanding of the mediators and kinetics of plant defense induction, and increasingly knowledge-based biological development and marketing strategies can be expected.

Continued focus on biopesticide formulation - especially the development of biogenic micro- and nanocapsules - will allow the inclusion of natural elicitors of plant defenses while facilitating the protection and timely release of biological agents and metabolites.

Local environmental conditions and requirements will drive targeted sourcing and positioning: Biologicals efficacy is heavily dependent on environmental conditions, and we should expect to see an increased focus on sourcing and positioning biologicals into local environments.

While developing countries are promoting the use of locally sourced biologicals through simplified registration, R&D grants and the support of local production units, it is expected that scouting for microbial candidates will also become more prevalent in EU regional environments and that EU commercial focus will extend from agriculture to horticulture and forestry.

2

STRATEGIC OPPORTUNITIES & CHALLENGES

By introducing unique modes of action, biopesticides can be used in integrated strategies with traditional pesticides to increase crop yields by working synergistically with chemical pesticides, to extend application timings and allow timely reuptake intervals as well as to provide resistance strategies.

R&D OPPORTUNITIES & CHALLENGES

Opportunities

Research centers around the world are undertaking research to improve techniques for the development and application of biopesticides to enhance their commercial feasibility.

Start-ups and other smaller, innovative companies have created a powerful and competitive environment for biopesticide R&D.

Acquisitions, collaboration- and licensing agreements bear witness to the ongoing symbiosis in the crop protection market, and the need for Inclusive Strategic R&D Management frameworks as well as the mediation of biology-based strategies to decision-makers is greater than ever before.

Currently, focus is on innovation at the interface between chemical and biological crop protection, precision technology and plant biology. Other technological developments include the integration of "big data" into pest management, including crowd-scouting by farmers and the increasing use of degree-day models for pest and pathogen prediction.

Ongoing legislation amendments to make the regulatory process appropriate for biopesticides facilitate the development and introduction of sustainable, non-chemical alternatives to pesticides, advancing the use of biopesticides in the EU.

Further harmonization between Member States will accelerate regional acceptance of biopesticidal strategies.

Challenges

Low financial incentives (return on investment) on the part of the incumbent crop protection industry restrict the adoption of minor-crop biopesticidal products. Furthermore, there is increasing awareness of natural plant toxins, which may be produced through the induction of plant defense responses by biopesticides.

Regional exemptions from residual tolerance levels hinder global trade to regions with maximum residue levels in food (such as the EU), reducing the potential economic viability of new biopesticides and thus their introduction into global crop protection strategies.

The cost of R&D registration and introduction can preclude small standalone companies entering the market. Start-ups and other smaller, innovative companies face further challenges in competing with incumbent agrochemical technologies, when attempting to convince risk-averse growers to transition to this emerging technology, but also with respect to the economic requirements for efficacy testing, registration and distribution.

Corporate IP policies further inhibit collaboration with agtech startups - incumbent corporations tend toward in-house development and innovation, often preferring acquisition rather than collaboration.

For corporations, acquisitions may lead to R&D funds becoming too diluted for them to prioritize true product innovation and strategic vision.

The inclusion of expert biological knowledge into R&D and business strategies is still not optimal. For current industry leaders in strategic growth and change, priority should continue to be given to developing Strategic R&D Management frameworks for leading innovation growth.

TECHNICAL OPPORTUNITIES & CHALLENGES

Opportunities

The trend in conventional and biopesticide formulation is to move away from wettable powders and suspension concentrates to water dispersible granules, and from dusts to granules.

Controlled release formulations are being developed to optimize effect, while nanotechnology is expected to provide new types of formulations such as nanoemulsions, nanosuspensions and nanocapsule formulations. Another area of focus is the choice of appropriate adjuvants for optimal biopesticidal activity.

Here, collaboration between biologists and chemists is facilitating the development and introduction of new formulations and applications, reducing the inconsistency of field performance and encouraging growers to implement this new technology.

Current research efforts in bioformulation technology are focused on delivery and application practices, especially on understanding the impact of plant defense induction kinetics on application timing and placement of the formulated product.

Challenges

Until recently, slow progress in formulation and delivery research has been a major hurdle to biopesticide development and acceptance.

There are significant costs involved in solving the biological issues of biopesticide formulations (including storage and field stability as well as efficacy consistency), and these can preclude small standalone companies entering the market.

However, formulation is one of the key factors in determining the commercial success of a biopesticide candidate, and the benefits in terms of market access and the acceptance of biopesticides as viable commercial alternatives or supplements to conventional plant protection products suffice to warrant this effort.

END-USER OPPORTUNITIES & CHALLENGES

Opportunities

The current model of relying almost exclusively on synthetic chemicals is changing.

Biopesticides present unique and complex modes of action and agronomic practices. Most agricultural biopesticides are suited for incorporation into IPM programs, in conjunction with traditional pesticides which, in turn, is contributing to the rapid increase in the use of biopesticides.

Today more and better biological active ingredients and products are available that can compete with, as well as complement conventional chemical pesticides. Demand, too, is driven by our increased knowledge about biopesticides mode of action and refined application rates and methods.

Challenges

Currently, biopesticides may demand higher unit prices compared to chemical pesticides, in part due to economies of scale, and the associated limited product availability is a further restraint to the growth in acceptance of biopesticides. On the other hand, locally produced biopesticides may be less costly than conventional, chemical pesticides.

There is a perception of lower and variable efficacy shorter persistence and greater dependence on environmental conditions on the part of end users.

Too often growers experience the frustration of inconsistent efficacy following the application of biologicals when disease pressures are already too high.

When applied with due regard for factors such as plant-response kinetics and disease pressure development, these same products prove to be highly efficacious.

Currently, reports of inconsistent efficacy lead to either over-optimism or skepticism on the part of marketers, corporate decision makers and farmers alike.

Biopesticides require knowledge-intensive management systems to develop and use them effectively. Crop protection biologists have a unique opportunity to mediate an understanding of the drivers of biopesticide efficacy, and to ensure innovation- and knowledge dissemination in the development and implementation of commercial strategies.

An increasing willingness to invest in field testing by agrochemical majors is addressing the perception of low efficacy and greater variability for biopesticides. However, start-ups and other smaller, innovative companies may not have funds for extensive demonstration trials and customer education.

3

FACTORS AFFECTING THE BIOPESTICIDE MARKET

A SWOT (strengths, weaknesses, opportunities, and threats) matrix (Figure 3.1) may be used to identify factors favorable and unfavorable to the development and application of biopesticides in crop protection strategies.

Strengths

A zero-day pre-harvest interval is of considerable benefit for producers of high-value crops, such as greenhouse vegetables, orchard crops and grapevines, by extending existing crop protection strategies to allow the control of late-season diseases.

Biopesticides are generally biodegradable, leave few (or no) harmful residues (easier residue management) and are less likely to harm non-target species.

Biopesticides may be incorporated in tank-mix or rotational programs with conventional pesticides (to obtain additive effects or synergies, or as part of a resistance strategy), or used as standalone applications.

Biopesticides may be effective at moderate to severe disease pressures and can be as effective as conventional chemical pesticides. Biopesticides can be cheaper than chemical pesticides when locally produced, and are currently quicker and cheaper to register than conventional pesticides.

STRENGTHS	WEAKNESSES
• Zero-day pre-harvest interval • Use under moderate to severe disease pressure • Tank mix / rotational program candidates • Quicker registration / organic farming • Can be cheaper than chemical pesticides • Can be more effective than chemical pesticides • Biodegradable; less likely to harm non-target species	• High specificity • Often slow speed of action • Often variable efficacy • Target resistance • Issues with lifespan & application technology
OPPORTUNITIES	**THREATS**
• Seed treatments and soil amendments. • Biostimulation against biotic & abiotic stresses • IPM; Mass production; fermentation • Formulation and application technology • New technology: growth potential	• Grower objection • Regulatory changes • Plant-defence toxins (current debate on toxins in organics) • Counterfeit CPP

Figure 3.1: SWOT matrix identifying factors favourable and unfavourable to the development and application of biopesticides in crop protection strategies.

Opportunities

Biofungicidal seed treatments are particularly suitable for the direct control of soil- and seed–borne pathogens. Furthermore, biopesticides may stimulate plant defense responses, which can increase the tolerance of treated crops to biotic as well as abiotic stresses.

Many biopesticides can be effectively produced using fermentation techniques and may be applied in a manner similar to chemical pesticides, thereby providing innovation opportunities within existing and new formulation and application technologies.

The growth potential of new biopesticide technologies is considerable - the market share of biopesticides is expected to rise from 2% of the global pesticide market in 2003 to 8% in 2020.

The increased public awareness of the benefits of *Integrated Pest Management* (IPM) strategies, which combine the use of biopesticides with natural predators, pest-resistant varieties, cultivation techniques and conventional crop protection chemistries.

An increased understanding of plant defense responses and plant–plant communications systems, as well as of abiotic and biotic factors, such as weather patterns and disease distribution kinetics, will allow us to develop early warning protective biocontrol strategies to communicate and work together with crops to reduce production losses.

• • •

Weaknesses

The high specificity of some biopesticides requires an advanced understanding of pest/pathogen biology and dynamics on the part of the user. Producers will increasingly need to provide this information to their commercial staff, as well as to distributors and customers.

Biopesticide efficacy is not only determined by the ability of microbial metabolites to directly control diseases and pests, but also by the induction of plant defense responses prior to pathogen attack. Plant defense responses are, however, not initiated until pathogens are detected, due to the energy costs of their induction and maintenance.

Biopesticides thus often have a slow speed of action, and have protective rather than curative activity, making them unsuitable if a pest or disease outbreak is an immediate threat to a crop.

Living organisms evolve and increase their resistance to biological and conventional control products, and strict anti-resistance measures must therefore be maintained.

Variable biopesticide efficacy (often cited as a key weakness) is in part due to the influences of various biotic and abiotic factors, and in part due to a lack of biological understanding of the effect of these influences on efficacy - more than ever, it is up to producers to mediate this information to their customers.

In addition, biopesticides require optimal storage facilities, a good understanding of the effect of storage on product viability as well as strict adherence to label recommendations to maintain the required shelf life and efficacy.

Threats

Adoption rates for biopesticides are currently low, as many growers still tend to prefer familiar conventional pesticides, in part due to reports of variable efficacy and slow speed of action for biopesticides.

Due to their slow speed of action, biopesticides are generally protective, rather than curative. Growers, on the other hand, tend to wait until the onset of symptoms before investing in crop protection, and thus require products with a certain level of curative activity.

The adoption of new agricultural practices - including biological pest control - by growers will depend on factors such as practicality, economics and political pressure, as well as on an understanding of the factors driving the efficacy of biopesticides.

The probable introduction of stricter EU and National regulations can be expected to restrict the growth of the biopesticide market. A threat in this respect is the increasing focus on natural plant toxins, which may be produced through the induction of plant defense responses by biopesticides.

4

BIOPESTICIDE DISCOVERY & DEVELOPMENT

AgChem companies are expanding their biopesticide R&D and portfolios, driven by stricter pesticide regulatory requirements, consumer awareness regarding hazards caused by chemical pesticides and the phasing out of toxic conventional pesticides.

Biorational process: the design and implementation of biopesticide discovery and development processes through screening (and the conservation of) specific biological effects (including pest and pathogen control, crop stress alleviation and enhanced plant resilience) against the pathogens and pests they are intended to control.

One of the key challenges to this process is the discovery and development of biopesticidal candidates.

Biopesticide discovery and development follows a *biorational process* of scouting / discovery and screening over lead development, formulation and scale-up, to application development, manufacturing and commercial introduction.

Biopesticide scouting

Biopesticide scouting - the identification of biopesticidally-active natural candidate species - includes observation in the field (the identification of plants showing exceptional natural growth or pest-defense responses) as well as ethnobotanical, anatomical and physiological indicators to identify potential candidates.

Ethnobotany: the scientific study of traditional knowledge and customs regarding plants and their uses.

Bioinsecticide discovery

Predating the development of synthetic pesticides during and after WWII, the ethnobotany of plant-derived pesticides - primarily insecticides - dates back at least two thousand years to Asia and the Mediterranean regions. In Europe and North America, plant-derived pesticides date back more than 150 years.

One of the most-cited ethnobotanical bioinsecticide examples is the use of plants of the genus Pyrethrum (now classified as Chrysanthemum or Tanacetum) in insecticide preparations, dating back to Persia about 400 BC.

Brought to Europe in the early 19th century, dried and ground flowers of Dalmatian chrysanthemum (*Tanacetum cinerariaefolium*) were used to control body lice during the Napoleonic Wars. Today, pyrethrum and its pyrethroid synthetic analogues are among the most widely-used biological and conventional insecticides.

Neem (*Azadirachta indica*) is an evergreen tree native to the Indian sub-continent, used for thousands of years against household and storage pests as well as pests related to field crops, and currently attracting global attention due to its potential as a commercial biopesticide.

Biofungicide discovery

In 1988, East German scientists tasked with the production of spores of the fungal plant pathogen, *Sclerotinia trifoliorum*, discovered that the pathogen was inhibited by a fungal parasite.

Recognizing its potential as a natural pesticide process, C. *minitans* was successfully developed as a commercial biopesticide by its discoverers, following the reunification of Germany.

The discovery that non-pathogenic microbes can induce plant defense responses is certainly not new. My grandfather famous for his roses which never seemed to be afflicted by mildew – would follow *his* grandfather's example, and spray yeast-rich beer dregs over his roses.

We now understand that the exposure of fungal yeast cells to plant leaves leads to the induction of plant defense responses, as the plants believe themselves to be under invasion by a potential fungal pathogen.

The EU pesticides database for Basic Substances (substances with a low toxicity profile not predominantly used for plant protection purposes but which could be useful in plant protection) includes Beer as one of the only 18 approved substances, while one of only 10 approved Low-risk Active Substances is Brewer's Yeast (*Saccharomyces cerevisiae)* - approved for use as a fungicide.

In a more recent example, Burkina Faso farmers are - together with researchers at Copenhagen University - testing the effects of a local plant (False Daisy, *Eclipta alba)* discovered to contain antifungal compounds, to develop inexpensive plant extracts capable of replacing synthetic fungicides.

Biofungicide discovery is currently being driven using biotech tools, including the monitoring of defense-related gene expression, pathogenesis-related proteins, bioinformatics (the collection, classification, storage, and analysis of biochemical and biological information using computers) and genomic analysis.

Bioherbicide discovery

Bioherbicide discovery typically involves the study of biochemical processes specific to plants, including photosynthesis, amino acid biosynthesis, chlorophyll and carotenoid biosynthesis and growth regulation (phytohormones).

Our growing understanding of these plant-specific processes is facilitating the identification of new herbicides sites of action, permitting the screening and development of bioherbicidally active metabolites with low mammalian toxicity, shorter environmental persistence, and novel target specificity.

Biopesticide lead development

In biopesticide lead development, microbes isolated from soil and plant microbiomes associated with plants showing exceptional growth or natural pest-defense responses are sequenced and compared with a database of known pathogens.

Matches are discarded, while the remainder transition to bioassay screening to identify biostimulatory or biopesticidal effects.

Potential leads are analyzed and characterized to ensure the absence of toxins, before embarking on a process of lab and field testing, fermentation process development, formulation and scale-up.

Each of the steps in this process of biorational development is subject to considerable challenges, but for biological pesticide development the transition from lab to field testing is one of the most crucial - correlation between a greenhouse result and a field result is not necessarily a given.

This is in part due to the complexity of biotic and abiotic factors affecting microbial characteristics, and the transient formulation stability of naturally derived metabolites and microbes.

The latter factor - rapid degradation by e.g. ultraviolet light, oxygen and microbial activity - contributes to shorter environmental persistence, a key factor driving the acceptance of biopesticides.

For conventional chemistries, development times can be three times longer than for biological crop protection development, while expenses may be as much as 10 times greater.

Industry figures suggest that the R&D process for a novel synthetic pesticide costs more than 250 million USD and requires 10 years of development, while the development of a biopesticide may cost 25 million USD and require three years of development.

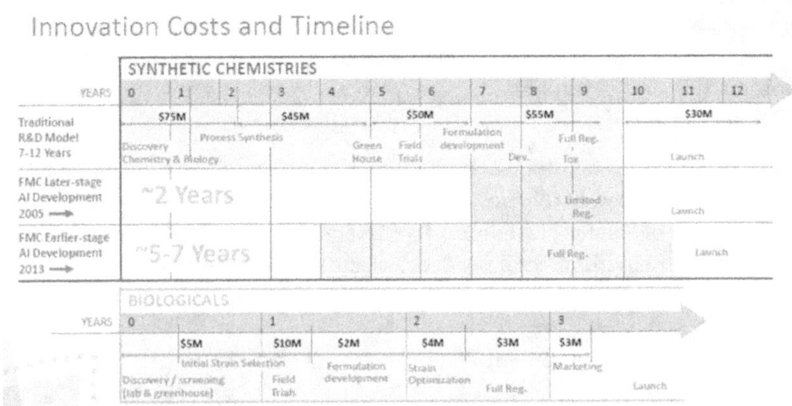

Figure 4.1: Innovation costs and timelines for conventional and biological crop protection development (from: FMC - Investor Relations Solutions, 2015).

Because of these differences, AgBio start-ups and other local, innovative companies have created a powerful and competitive environment for biopesticide discovery and development, and the current wave of acquisitions, collaboration- and licensing agreements with the incumbent AgChem industry bear witness to an ongoing symbiosis in the field of biopesticide R&D.

5

STRATEGIC VISION FOR THE CROP PROTECTION INDUSTRY

The crop protection industry faces a dilemma.

In the uncertain and globally competitive world of crop protection, focus has primarily been placed on incremental product development, rather than on strategic vision and ideation.

Because of this, the incumbent crop protection industry risks stagnating. Competitive cost measures trump strategic vision.

The State of our Industry

To understand our dilemma, it is necessary to know that while the post-war era of discovery and commercialization was a period of continuous innovation, no major new mode of action has been introduced to the market in the past 30 years.

Since the 1980s, crop protection corporations have sought to grow through global expansion and incremental, "good-enough" product development, relying on revenues from successful chemicals that were decades old.

Even commercial biopesticides were primarily based on natural mechanisms developed over more than 100 million years.

Neither product innovation — based on long-term research projects with uncertain outcomes — nor strategic vision were primary concerns. Competitive cost structure was.

. . .

The best of times, the worst of times

With the financial crisis of 2008, globalization halted, and crop protection products became commodities. Since then, corporations have been expanding through acquisition, becoming larger and more competitive, but also more cumbersome.

For many corporations, R&D funds are becoming too diluted for them to prioritize true product innovation and strategic vision.

In this climate of competitive cost structure and "growth" by corporate acquisition, the coming-of-age of the biopesticide industry and its inclusion in integrated crop protection strategies is seen by many as a sign of a new period of innovation.

Although start-ups and other smaller, innovative companies have created a powerful and competitive environment for biopesticide R&D, these small companies face considerable challenges as a result of acquisitions, collaborations, and licensing agreements with the incumbent chemical industry.

While these bear witness to the ongoing symbiosis in the crop protection market, the need for ideation, innovation and vision across the industry is greater than ever before.

In the uncertain and globally competitive world of crop protection, organizations not prioritizing strategic vision and change will be supplanted by those that do.

I believe that, like the more agile computer industry, the crop protection industry — conventional as well as biological — must continue to shift focus from "product" to "implementation".

Within the industry, the strategic shift from "product" to "implementation" will rely on Strategic Ideation to identify powerful ideas for growth.

Successfully executing this change will depend on our ability to not only integrate synergistic conventional, biological and technological solutions, but also on our ability to facilitate and integrate intelligent communication with crops.

Innovation interfaces

Currently, the biggest innovations in the crop protection industry are at the interface between chemical and biological crop protection, precision technology and plant biology.

I envision a future where growers will protect their harvest in close cooperation with the crops themselves, monitoring weather and disease patterns and communicating with crops through complex systems developed over millennia: leaf- and root receptors, volatile stress hormones, root-root and root-mycorrhiza interactions and inducible, systemic defense response cascades.

> *"I think the biggest innovations of the 21st century will be at the intersection of biology and technology. A new era is beginning."*
>
> - STEVE JOBS (WALTER ISAACSON)

In less than 10 years, I foresee precision crop-signaling products strategically placed in fields, able to be automatically released and activated by weather- and pest-monitoring systems.

Local, induced plant defense responses will be transmitted throughout the field by inter-plant communications, proactively priming crops to combat invasive pests and pathogens.

"Blanket" applications of chemical and biological crop protection products will be eliminated, and aerial monitoring as well as precision application solutions will permit site-specific supplemental treatment, when and as required.

Industry Leaders

For current industry leaders in strategic growth and change, priority is being given to developing Strategic R&D Management frameworks for leading innovation growth.

These corporations obtain insight into change initiatives as well as cultural and leadership challenges within their own organization, such as transitioning strategic thinking to entrepreneurial thinking by introducing entrepreneurial mindsets and expert intuition in established R&D frameworks.

Significant opportunities outside traditional, vertically-integrated industry roles exist for industry experts able to provide ideation as well as strategic and technical expertise to both start-ups and existing corporations.

The strategic shift from "product" to "implementation" will only be successful if we prioritize a deep technical understanding of pesticide and biopesticide mode of action and formulation, precision scouting and application technology, as well as the essential principles of plant and pest biology.

II

STRATEGIC RESEARCH & DEVELOPMENT

6

R&D IN THE GLOBAL CROP PROTECTION INDUSTRY

In scientific or technological organizations, the primary function of Research and Development (R&D) is to develop new products by driving future product development.

> "It's really hard to design products by focus groups. A lot of times, people don't know what they want until you show it to them."
>
> - STEVE JOBS

In industrial scientific R&D, "Research" may be defined as a process addressing the questions "what is the challenge, and what is its solution?" while "Development" may be defined as a process addressing the question "how do we utilize the solution to the challenge in a commercially viable manner?".

For the crop protection industry, "Research" may include processes addressing the challenges of pesticide resistance as well as the commercial need for the introduction of new products, through the discovery of pesticidal active ingredients with new or improved modes of action.

"Development" may be described as the processes addressing the optimization of active ingredient efficacy, safety and commercial viability.

. . .

R&D Portfolios

The market position of a company (incumbent or new) determines - to a large extent - their R&D and Innovation strategies.

For mature corporations and traditional businesses with a near monopoly, existing products and strategies may limit development, and these companies tend to focus on Defensive Core Portfolios.

Figure 6.1: Strategic R&D portfolios.

A concern for these companies is that focus on an aggressive growth portfolio and neglect of the core portfolio may lead to the competition using the opportunity to attack their core business.

New companies and start-ups which do not have a core business to defend do not have the same limits, and tend to pursue Aggressive Growth Portfolios.

R&D in the Crop Protection Industry

While the post-war era of discovery and commercialization was a period of continuous innovation, since the 1980s crop protection corporations have increasingly counted on growth through global expansion and reliance on revenues from existing products.

With the financial crisis of 2008, globalization stagnated, crop protection products became commodities, and corporations focused on "growth" through acquisitions, collaborations, and licensing agreements.

Many of these acquisitions and consolidations were driven by the need to spread the high fixed costs of pesticide R&D (specifically the cost of discovery and development of new active ingredients), production, marketing and distribution over a greater sales volume in more markets (economy of scale).

Within the global crop protection industry, for many companies R&D resources have become too diluted to implement true product innovation and strategic vision, and larger companies have leveraged consolidation to access the research and development initiatives of smaller, more innovative companies.

Recently however, R&D restructuring and investment by leaders in the crop protection industry has shown an increase, driven by technological developments such as the introduction of novel active ingredients, the adoption of genetically modified crops, new application techniques, formulation developments and the introduction of unique biopesticidal product ranges.

In this context, industrial R&D utilizes future-oriented long-term strategies common to scientific research, while retaining focus on commercial yield.

R&D is crucial for the development and introduction of new products, or for the incremental improvement of existing product ranges.

In contrast to most other corporate activities, R&D does not aim to yield immediate profit, and is associated with considerable risk and an uncertain return on investment.

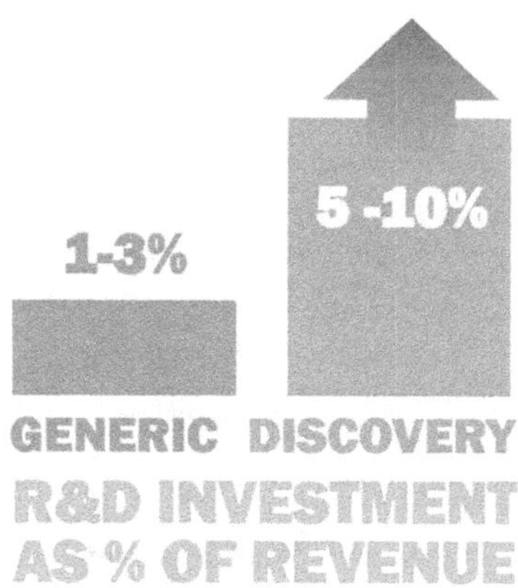

Figure 6.2: R&D spending ratios as a percent of revenues for crop protection product development.

Measures of R&D reflect the state of the industry, such as the level of competition or the degree of technological progress.

For the pesticide industry, R&D spending ranges from the low single figures for generic plant protection companies to more than 10% of total sales turnover (revenues) for discovery-based companies.

R&D spending ratios of more than 10-15% of revenues are unusual, representing a level more generally seen in high-technology engineering companies.

R&D processes in crop protection product development

For crop protection R&D, the process of identifying a novel active ingredient and converting it into a product that farmers can use can take more than 10 years, and requires an investment of more than €200 million (Figure 6.3).

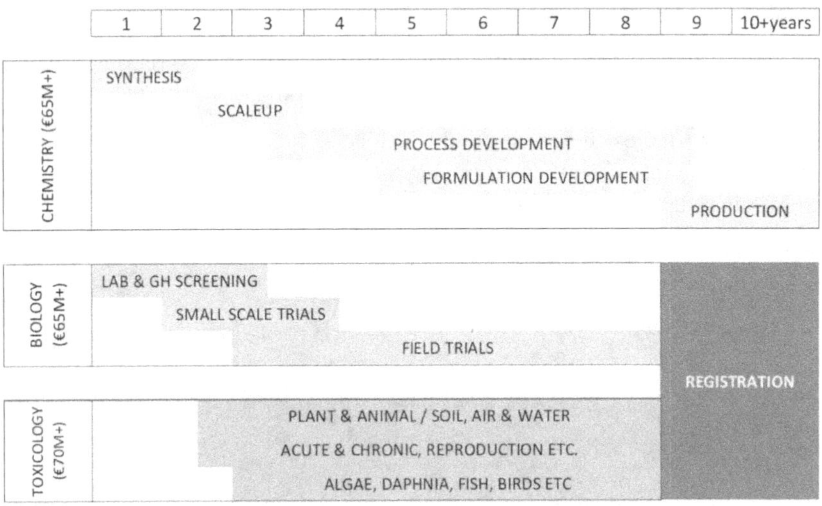

Adapted from: Innovation: the secret of our success (BASF, 2018)

Figure 6.3: The research and development process for crop protection products.

The process encompasses Chemistry R&D: including active ingredient research and formulation development, Biology R&D: including mode of action research and application development, and Toxicology: the determination of health and environmental risks.

For the crop protection industry, R&D identifies and addresses the challenge of how to enable proactive responses to technology change, competition and changing customer needs.

In the rapidly changing global crop protection landscape, organizations survive by continually improving their range of products.

R&D strategies in the Crop Protection Industry

For the crop protection industry, R&D strategies have typically included new product discovery (new active ingredients and technologies) as well as product development - including the development of off-patent products (generic products) and the development of new or improved products based on existing active ingredient portfolios (proprietary products).

New product discovery and development focuses on identifying and developing active ingredients with novel modes of action that can be combined with existing active ingredients in resistance strategies, that may provide increased effectiveness in specific crop species, that may offer a lower cost of production, and which may be patented.

Intellectual property protection, such as patents, give these companies the chance to recover the costs of discovery R&D, by providing competitive advantages such as posing a legal deterrent to imitators while offering first-mover advantages such as higher prices and profit margins.

New product development requires considerable R&D investment, with discovery companies typically spending 10% or more of their revenue on R&D. In contrast companies focusing on generic development typically spend 1-3% of their revenue on R&D.

Following the recent rounds of acquisitions, collaborations and licensing agreements, increased focus is being placed on R&D investment in the integration of biological and chemical technologies for crop protection.

The rise of "integrated" crop protection companies is a major development that has lead to consolidation among agricultural seed and -chemical companies, and the introduction of genetically modified seed varieties (GM crops) with complementary chemical crop protection products.

Examples include crops engineered to be resistant to the herbicides glyphosate and glufosinate, and the development of "plant-incorporated protectants" - a form of biological insecticide in which foreign DNA such as the BT-toxin gene is inserted into crop genetic material.

R&D Trends in Crop Protection

Due to the technical challenges and high costs related to the successful discovery and development of active ingredients with novel modes of action, R&D investment by crop protection companies is shifting away from research (discovery) back to development.

Renewed focus is being placed on increasing the effectiveness and safety of existing active ingredients and formulations, by optimizing pesticide application and uptake, by decreasing the persistence of chemicals in the environment and by reducing toxicity to humans and other non-target species.

While increasing R&D and regulatory investments are driving the development of crop protection products with fewer health and environmental risks, there is a tendency for the industry to focus on products on the basis of their commercial potential - at the expense of products for smaller markets.

A separate development trend in the crop protection industry is the increasing investment in biopesticide products and strategies.

Biopesticides are naturally occurring biological agents, substances or mixtures of substances used to prevent, destroy or control pests (including weeds) and diseases, and are currently exhibiting strong growth in the global pesticide market.

R&D investments in biopesticide product and strategy development focus on improving techniques for the development and application of biopesticides to enhance their commercial feasibility.

7

STRATEGIC R&D MANAGEMENT

In the crop protection industry, Research and Development (R&D) of new active ingredients, optimization of product portfolios and the introduction of innovative crop protection strategies are a necessity to ensure the provision of sufficient crop resources to meet the increasing needs of the world's population, as well as enabling proactive responses to technology change, competition and changing customer demands.

> "You can't just ask customers what they want and then try to give that to them. By the time you get it built, they'll want something new.""
>
> \- STEVE JOBS

Strategic R&D Management prioritizes the integration of R&D with corporate and business objectives, while ensuring compliance with conventional Project Management parameters, such as budgets and performance indicators.

R&D Management

R&D management enables organizations to respond to external or internal opportunities, and to introduce new strategies, processes or products.

Figure 7.1: Summary of R&D management activities.

R&D management may be described as the process of designing R&D strategies, managing technical expertise (creation and retention) and innovation (creation and commercialization), the leading of talent teams, and the transfer of knowledge and innovation within the company or network (Figure 7.1).

Strategic R&D management activities may include basic and applied research management, technical-, product - and process development management, talent management and strategic management, as well as IP- and innovation management.

R&D Management vs. Project Management

One of the main objectives of R&D management is to ensure that projects are completed successfully and within the defined timeframes and budgets - a basic tenet of conventional Project Management.

As for any management system, basic parameters such as budget- and performance requirements need to be managed with an appropriate level of discipline.

This is especially true for managing R&D teams and projects, where scientists may have a tendency to focus on the process more than on the needs of the customer.

However, due to the nature and complexity of R&D, and the need to provide an environment conducive to ongoing iteration (the implementation of experimental learning into the redesign of new trials) and innovation, the rigid application of conventional project management frameworks risks creating barriers to innovation.

For R&D projects, dedicated R&D management models are needed, capable of adhering to the need for efficiency and discipline while ensuring sufficient freedom for innovation processes.

R&D and Strategic R&D Management Models

R&D and Strategic R&D Management models may be described in terms of maturity: so-called first, second and third generation R&D Management models have the objective of creating and disseminating scientific knowledge and innovation internally *within* the organization (endogenous focus) while fourth generation models introduce the concept of globalized generation of intellectual capital through collaboration *across* organizations (exogenous focus).

In *first-generation R&D Management models,* the objective is to facilitate scientific discovery with emphasis on the scientific intuition of the individual researchers. Project funding and prioritization are governed by perceived scientific value (science "push") and budget, rather than strategic parameters (market "pull").

In these systems, researchers may operate with considerable autonomy and first-generation R&D Management models are often characterized by flexibility in timelines, budgets and prioritization - often with low accountability.

Second-generation R&D Management models may be considered an extension of first-generation management models, and adopt a more market-orientated strategy for prioritization.

Standard Project Management frameworks (such as key performance indicators, output-based KPIs, and binary "gated" decision-making processes) are typically used on a project-by-project basis to ensure accountability and compliance with budgets and timelines.

With the emergence of increasingly large crop protection corporations, the need arose for strategic management models capable of facilitating the interaction between R&D and Business units, in order to ensure a strategically balanced product portfolio.

R&D Management Generation	Objective	Focus	Drivers	Management strategy
4th Generation	Generation of intellectual capital through **collaboration**	**Exogenous**	**Collaborative** innovation	**Network orientated:** Strategic Innovation & Knowledge Management
3rd Generation			**Business** strategy	**Portfolio orientated:** Portfolio- and Strategic R&D Management
2nd Generation	Creation & dissemination of innovation **within the organization**	**Endogenous**	**Market "pull"**	**Project orientated:** Project management
1st Generation			**Science "push"**	**Researcher orientated:** flexible timelines and budgets

Figure 7.2 : Summary of R&D management models.

Accordingly, *third-generation R&D Management models* are portfolio orientated, driven by business strategy, and prioritize the integration of R&D with corporate and business objectives.

To ensure compliance with strategic objectives, dedicated R&D Management techniques (such as Iterative Gate processes and Inclusive Management Frameworks) are typically implemented.

Increasing globalization, including acquisitions, collaborations, and licensing agreements, challenge existing R&D Management models with respect to exogenous activities outside the organization. In the current consolidation climate, significant opportunities arise outside traditional endogenous company restraints.

In a world of globally distributed knowledge, companies are no longer able to rely entirely on their own research, but are increasingly dependent on networking as a source of innovation.

Global acquisitions, collaboration, and licensing agreements increasingly require global R&D Management strategies, which may be described as the process of designing R&D strategies and processes across multi-cultural and multi-lingual settings.

To utilize these opportunities, global corporations are increasingly making use of *fourth-generation R&D management models,* with the objective of generating intellectual capital through external collaborations. These fourth-generation models are network-orientated and driven by collaborative innovation.

In addition to existing Project Management and Strategic R&D Management techniques, strategic innovation and knowledge management methods may be implemented to ensure the synergistic integration of R&D strategies across organizations, cultures and fields of scientific endeavor.

For globally orientated organizations, fourth-generation R&D Management models utilize the advantages of collaborative innovation, including synergism between internal and external R&D teams, the reduced cost of conducting research and development and shared market research and market reach.

In addition to the need to train scientists and managers in cross-cultural communication, fourth-generation models and the globalization of R&D Management may require additional considerations, such as:

- the post-merger integration (or dissolution) of R&D groups as well as decisions regarding their location or relocation
- the realignment of innovation strategies
- the introduction of R&D networks and virtual R&D Management systems
- the transfer of technology and the sharing (and retention) of expertise, and
- the introduction of cross-cultural R&D innovation- and performance index parameters

All the described R&D management models have the objective of improving R&D performance and productivity, providing structure to the R&D process, and ensuring an R&D environment in which the complexity of R&D can be managed.

Through the application of dedicated management models, R&D Management is crucial for the design of R&D strategies, the management of technical expertise and innovation, the leading of talent teams, and the transfer of knowledge and innovation within the company or network.

———

8

"BLUE OCEANS" OF UNCONTESTED CROP PROTECTION MARKET SPACE

Since the 1980s, crop protection corporations have sought to grow through global expansion and incremental, "good-enough" product development, relying on revenues from successful chemicals that are decades old.

Today, organizations not prioritizing strategic vision and change risk being supplanted by those that are able to create market-shaping innovations.

Successfully executing strategic vision will depend on our ability to not only integrate synergistic conventional, biological and technological solutions but also on our ability to facilitate and integrate intelligent communication with crops.

The need for strategically balanced portfolios

In the globally competitive world of crop protection, long-term survival is increasingly dependent on a shift from conservative product portfolios ("Settler" and "Migrators") to strategically balanced portfolios which include "Pioneer" products.

"Settler" products comprise the mainstay of generic portfolios, and the term is used to describe generic or copied products, for which the main competitive driver is cost domination.

"Migrators" are products with value improvements, typically in the form of formulation developments, for which the main competitive driver is improved efficiency.

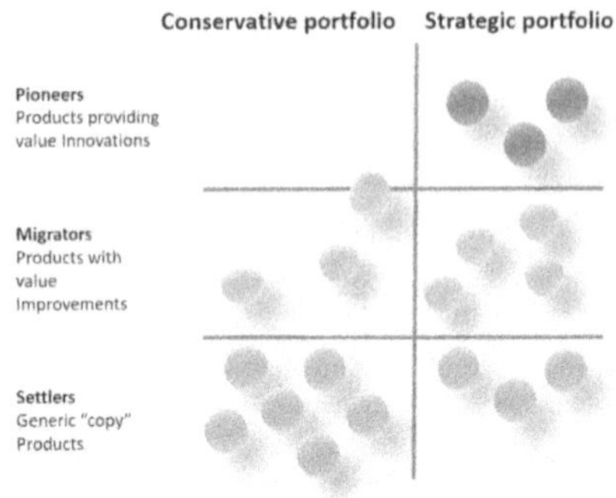

Figure 8.1: Pioneer-migrator-settler map for the prediction of future growth and profit (adapted from Kim & Mauborgne, authors/creators of Blue Ocean Strategy).

"Pioneers" are products providing value innovations, such as the integration of technical advancements or novel technologies.

Driving Value Innovation

Crop Protection corporations do not need to choose between a generic strategy of cost domination or differentiation but may simultaneously pursue differentiation and low cost (Value Innovation).

Value innovation may be defined as the simultaneous creation of new utilities (values) at a lower cost. Costs may be reduced by eliminating conventional competition factors wild value may be raised by creating new competition factors.

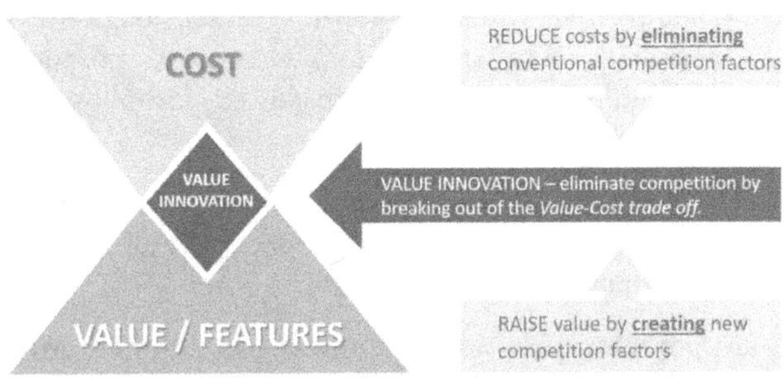

Figure 8.2: Value innovation through the elimination of conventional - and the creation of new - competition factors (adapted from Kim & Mauborgne, authors/creators of Blue Ocean Strategy).

In value innovation, uncontested market spaces may be created by breaking out of the value / cost trade-off.

Identifying uncontested market spaces (Blue Oceans)

Through value innovation, competition is rendered irrelevant by opening new and uncontested market spaces (Blue Oceans).

Blue Ocean Strategy is a theory of market creation developed by W. C. Kim and R. Mauborgne, Professors of Strategy at INSEAD, one of the world's top business schools, and co-directors of the INSEAD Blue Ocean Strategy Institute in Fontainebleau, France.

Blue Ocean Strategy proposes that companies can eliminate competition by creating "blue oceans" of uncontested market space, primarily based on value innovation and in which competition is essentially eliminated, as opposed to "red oceans" - characterized by primarily incremental innovation and limited growth prospects.

To identify value innovations for the crop protection industry, Blue Ocean Strategy recommends that organizations create value innovation utility spaces by focusing on buyer (grower) product experience and pain points.

BLUE OCEAN	RED OCEAN
• Create uncontested market space	• Compete in existing market space
• Make the competition irrelevant	• Beat the competition
• Create and capture new demand	• Exploit existing demand
• Break the value-cost trade-off	• Make the value cost trade-off
• Pursue low cost ***and*** differentiation	• Pursue low cost ***or*** differentiation

Figure 8.3: "Blue Oceans" offer uncontested market space, while "Red Oceans" provide limited growth prospects (adapted from Kim & Mauborgne, authors/creators of Blue Ocean Strategy).

A Buyer (Grower) Utility Map can facilitate the identification of utility spaces that a product or service can fill, thereby contributing to new value propositions distinct from existing competition parameters.

Grower Utility Map

To create a Grower Utility map for the Crop Protection industry, grower experiences can be listed as sequential Grower Experience stages, for example:

- purchase of the crop protection product or utility
- scouting (identifying pest or pathogen levels and optimum application timing)
- application (delivery of the product to its target site)
- supplements (identifying and using the optimal adjuvant or co-product)
- effect (evaluation of sufficient effect to combat yield loss and reduce the risk of resistance development)
- crop response (elimination of crop phytotoxicity and induction of crop defense responses)

These stages may be broken down into Grower Utility levels, through which companies can provide utility for their customers. The most commonly exploited level is that of Grower Benefits, in which innovation can increase growers' productivity by increasing their efficiency. By evaluating the 36 spaces of the Grower Utility map, it is possible to identify utility propositions that are different from existing products.

Within the Crop Protection industry, most generic crop protection companies focus on one or two stages of the grower experience sequence (e.g. purchase and application effectivity), overlooking the opportunities offered by the remaining stages.

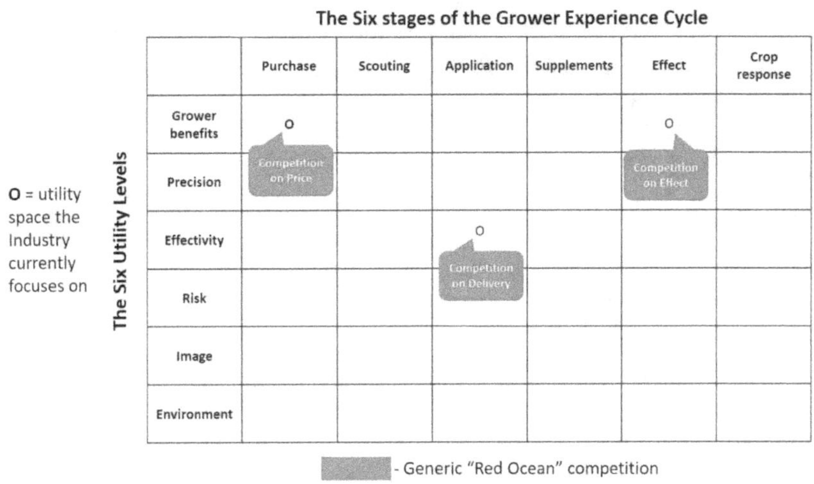

Figure 8.4: Grower Utility Map. The generic crop protection industry currently competes in a "Red Ocean" of "Settler "products (adapted from Kim & Mauborgne, authors/creators of Blue Ocean Strategy).

Focus has traditionally been placed on delivering more of the same utility at the same stage of the grower's experience, with innovation incrementally achieved through improvements in current utility propositions ("Settler and "Migrator" products, see Figure 8.1).

The current "state of play" of the competition's product strategy can be visualized and compared to one's own product strategy through a Strategy Canvas.

Strategy Canvas

The Strategy Canvas is a useful visual tool to compare a company's strategic profile with that of the competition. In the following example, the current strategic profile of a traditional generic crop protection company is compared to that of the generic crop protection industry.

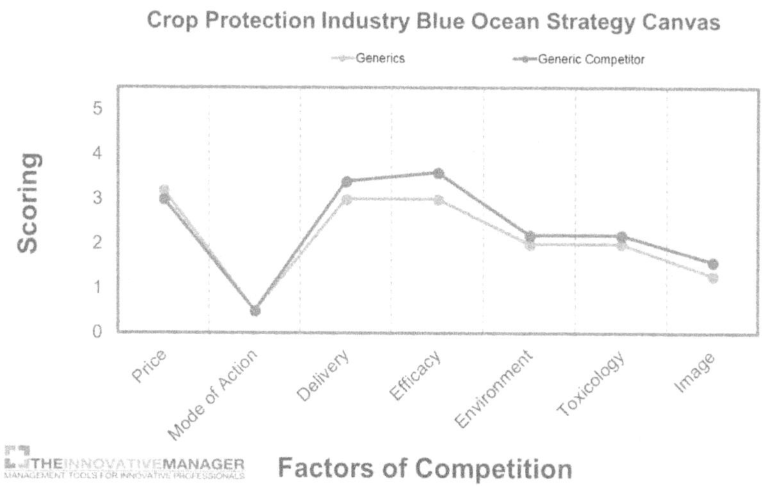

Figure 8.5: Strategy Canvas, generic crop protection competitor compared to the generic crop protection industry (adapted from Kim & Mauborgne, authors/creators of Blue Ocean Strategy).

In this canvas, the factors that the industry competes on are listed on the x-axis, and the priority across these factors is scored on the y-axis. The strategic profile of both the company and the generic industry can be seen to be almost identical. The traditional approach to generic crop protection product development has been to compete on prices and incrementally improve formulation and delivery technologies. This strategic profile is unlikely to create market-shaping innovations.

. . .

Advances in Biological Crop Protection provide value innovation opportunities and strategic adaptations

In contrast, the nascent Biopesticide industry staked out new utility propositions on the Grower Utility map, to generate a unique strategic profile.

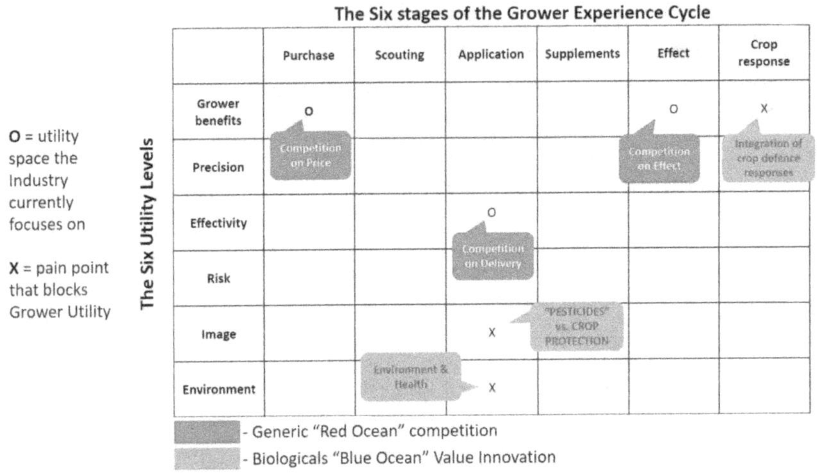

Figure 8.6: Grower Utility Map. The Biopesticide industry has created a "Blue Ocean" of "Pioneer" products and strategies (adapted from Kim & Mauborgne, authors/creators of Blue Ocean Strategy).

Over the last two decades, significant increases in our understanding of inducible plant defense responses (strongly driven by Biopesticide R&D).

This increased understanding provides us with new and powerful utility propositions to increase product efficacy, through the introduction of new modes of activity, and the integration of plant defense responses into overall crop protection strategies - an alliance between farmer and crop.

To generate their unique strategic profile, the Biopesticide industry altered their scoring on competitive factors through:

Elimination of competition on price (growers accept the increased price of Biopesticides)

Reduction of competition on delivery (application of Biopesticides is more demanding than for conventional, chemical products) and efficacy (there is a perception of variable efficacy for Biopesticides)

Elevation above industry standards of competition on new modes of action, environmental and toxicological advantages as well as positive branding (image)

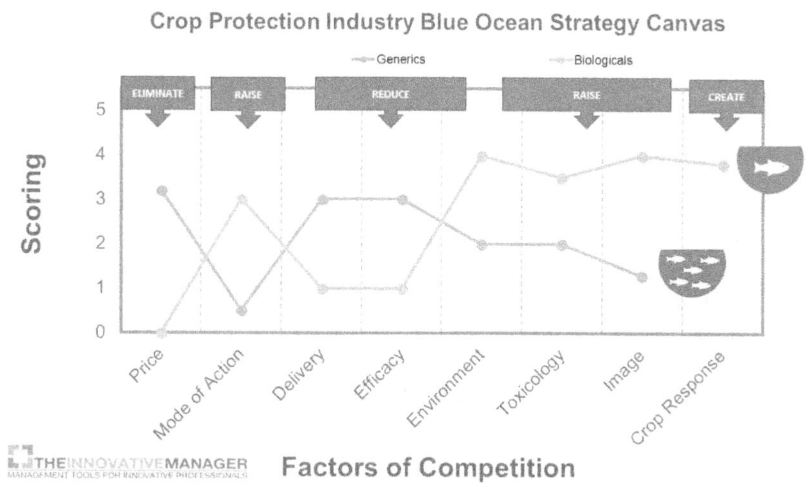

Figure 8.7: Strategy Canvas, Biopesticide industry compared to the generic crop protection industry. (adapted from Kim & Mauborgne, authors/creators of Blue Ocean Strategy).

... and generated "Blue Ocean" value innovation in an uncontested market space through:

Creation of a new factor that the incumbent generic industry did not offer: the integration of plant defense responses into overall crop protection strategies.

The Biopesticide industry has subsequently demonstrated that significant strategic synergies can be achieved by integrating biological and conventional crop protection - not only between products but also between the knowledge bases for these strategies.

While the generic product development strategy has been to incrementally improve formulation and delivery technologies, significant increases in our understanding of inducible plant defense responses (strongly driven by decades of biologicals R&D) provide us with a new and powerful tool to increase product efficacy, through the integration of plant defense responses into overall crop protection strategies - an alliance between farmer and crop.

Inducible plant defense responses have developed over more than 100 million years and have consistently and effectively ensured species' survival.

For integrated crop:product strategies in monocultures, however, the kinetics of plant defense induction require precise and timely activation - after confirmation of pest or disease presence, but before disease or pest pressures exceed a specific threshold.

A key success factor for the above is understanding - and the timely induction of - plant defense kinetics.

Fortunately, scouting and threshold monitoring through drone-based technologies are now available to provide the real-time information to make inducible plant defense responses a feasible option for the development of a new generation of integrated crop:product strategies.

SMART-TECH connects with Biological Advances to increase "Blue Ocean" Value Innovation

In response to this new generation of integrated crop: product strategies, the SMART-TECH industry has staked out new utility propositions on the Grower Utility map by building on the BioPesticide industries value proposition, to generate a unique strategic profile.

A key R&D and marketing objective for most conventional Crop Protection organizations has been to improve the efficacy of existing products, and thereby reduce the amount of active ingredient required to achieve optimal crop protection.

However, for true sustainability, it's no longer enough to look BACKWARDS and increase efficacy and reduce rates in response to regulatory and public pressure. In the SMART REVOLUTION, real-time scouting and information technologies will allow the timely inclusion of inducible plant defense responses for crop protection solutions.

This will drive rate and volume reduction as FORWARD strategies, such as targeted drone delivery, removing one of the major technical hurdles to making this leap forward toward truly sustainable Crop Protection.

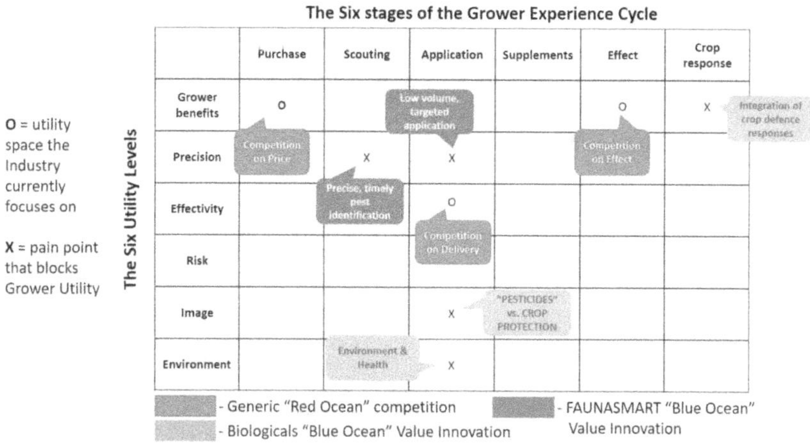

Figure 8.8: Grower Utility Map. The SMART-TECH industry has created a "Blue Ocean" of "Pioneer " products (products providing value innovations, such as the integration of technical advancements or novel technologies) and strategies, through advances in agricultural, satellite and drone technology and by building on the BioPesticide industries value proposition, to generate a unique strategic profile (adapted from Kim & Mauborgne, authors/creators of Blue Ocean Strategy).

The Danish startup FaunaSmart is an example of a new company contributing to the SMART REVOLUTION and is the first precision agriculture company where precision is also based on DNA, providing opportunities for sustainable, Integrated Pest Management.

FaunaSmart achieves its value innovation through the use of drones for monitoring, field sampling and real-time DNA identification of insect colonies in their early developmental stages.

In the FaunaSmart system drones are used as tools for real-time field inspection of pest species present. These drones offer accuracy and efficiency, analyzing fields with cameras or need-be DNA samplers.

FaunaSmart software analyses pest risk levels, provides real-time, in-field maps and when it displays different zones the software finds the most efficient and environmentally friendly solutions to control pests.

Together with formulation and delivery advances, real-time drone scouting and analysis will drive improved product efficacy - permitting significant reductions in product volume. This in turn will finally allow the development of targeted drone-based application technologies, for which a major technical impedance has been the requirement for low spray volumes.

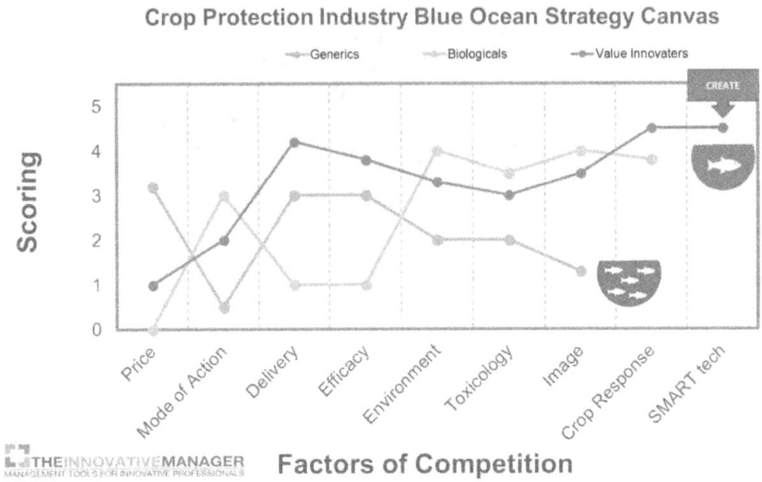

Figure 8.9: Strategy Canvas, the SMART-TECH and Biopesticide industries compared to the generic crop protection industry. "Blue Ocean" Value Innovation opportunities and strategic adaptations facilitate the identification and development of "Pioneer" products (adapted from Kim & Mauborgne, authors/creators of Blue Ocean Strategy).

A new generation of integrated crop:product strategies coupled with real-time scouting, monitoring and data evaluation as well as targeted drone-based application technologies can provide significant and sustainable opportunities for differentiation in the highly competitive crop protection market.

Executing Blue Ocean Strategies

Often, innovations need to overcome resistance from stakeholders within the company. Key organizational hurdles include cognitive, resource, motivational and political hurdles that prevent strategy executors from committing to the strategic shift, providing resources, leading innovation teams, and from overcoming external opposition.

III

STRATEGIC R&D MANAGEMENT

9

LEADING INNOVATION IN R&D TEAMS

To successfully manage the *critical interface between strategy (vision) & execution* for innovation portfolios, it is necessary to introduce an iterative, input-based and inclusive development framework to promote the inclusion and buy-in of expert teams.

How do we ensure a speak-up culture?

Traditionally, crop protection companies have implemented a classical *management career track*. In this system, competent R&D scientists are promoted to management positions, and it is from these positions that contributions are made to business strategy.

Unfortunately, it is not always the case that the most competent experts become the best managers or strategists, and it is quite common for excellent scientists to attain management positions and subsequently leave the company to return to the bench.

Alternatively, talented scientists may refuse promotion to managerial positions, and their input to strategic decisions is lost.

The introduction of an *expert-track career track* is a way to provide experienced and talented scientists with the opportunity to contribute to business strategy at a managerial level, without the constraints of a managerial career track.

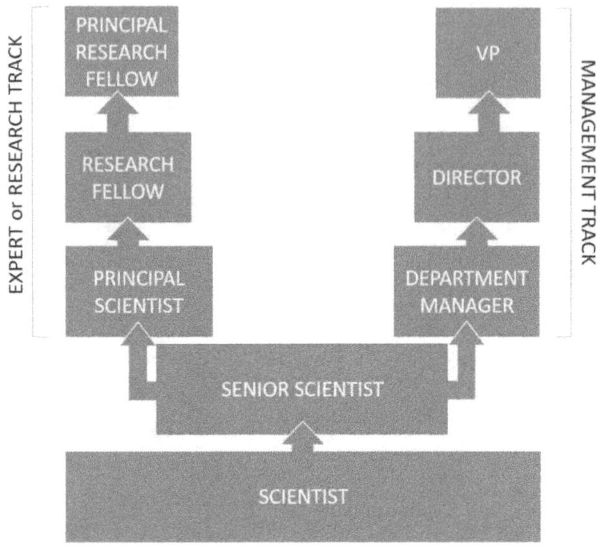

Figure 9.1: Dual career-track system

Here, talented scientists progress through Senior- and Principal Scientist levels to the position of Research Fellow, at which level they have the same opportunity to provide strategic input as *e.g.* an R&D Director.

How do we promote bottom-up initiatives?

Conventional, output-based incremental development frameworks are appropriate for the management of incremental (conventional) development projects, and output-based measures of performance such as KPIs are easy to define.

However, incremental frameworks may hinder the iterative R&D process of innovation, and KPI's and bonus incentives often prove detrimental to the innovation process. Simply put, bonuses do not always motivate scientists.

For innovation projects, *measures of input-based performance* can be more appropriate, although the identification and quantification of these performance measures prove challenging for many companies.

An example of an appropriate measure could be the quantification and recognition of learning provided to the project or company by the individual.

For input-based innovation projects, one objective of measuring performance is to provide a means to *end* the innovation process (*e.g.*, by defining a *minimum viable product*).

How do we lead and motivate expert teams?

Innovation and inclusion of expertise from the "execution" or "down" organizational layer entails recognizing and legitimizing innovation and learning inputs as a measure of input-based performance.

At the same time, clear expectations of independent, individual innovation should be included in R&D job descriptions.

> "It doesn't make sense to hire smart people and then tell them what to do, we hire smart people so they can tell us what to do."
>
> - STEVE JOBS

R&D experts should always have the freedom to challenge the existing R&D framework, to ensure "bottom-up" integration, for example by critically evaluating project-group makeup - "is Biology adequately represented throughout the project time plan?" and "do evaluation criteria also measure learning?".

Rather than monetary bonuses, R&D staff may be better motivated through *e.g.* the provision of time for them to pursue non-project related innovation initiatives, opportunities for scientific publication and increased access to scientific congresses and conferences not directly relevant for the company's core business.

Increased flexibility in working hours and locations for R&D staff as well as increased autonomy have also proven to be effective motivators.

Finally, the visible inclusion of R&D input into strategic business decisions by encouraging a speak-up culture and promoting bottom-up initiatives is a key factor in ensuring the motivation of expert teams.

10

MANAGING INNOVATION PROCESSES IN R&D

Innovation can be defined as the identification of new solutions (products and applications, business strategies and organizational innovations), or as the identification of improved solutions to adapt to changing requirements or markets.

> "Innovation has nothing to do with how many R&D dollars you have. When Apple came up with the Mac, IBM was spending at least 100 times more on R&D. It's not about money. It's about the people you have, how you're led, and how much you get it."
>
> - STEVE JOBS

Innovation is not restricted to R&D, nor to management levels within the organization, and requires participation across the organizational matrix (*e.g.*, product development, manufacturing and marketing) and at every level (from corporate management to expert), in order to ensure the generation of new solutions.

Innovation typically begins with the generation of ideas which may be formulated as hypotheses, able to be tested and evaluated.

Innovation may be driven by producer "push" (based on existing or novel technologies, for which newer, more profitable applications are developed) and market "pull" (areas where customer needs are addressed and solutions developed).

For both drivers, an understanding of both market and technical issues are needed, requiring multi-functional development teams comprising both commercial and technical participation.

Increased competition constantly reduces the product lifecycle of products, forcing companies to reduce time-to-market.

As a consequence, Innovation managers seek to decrease development time, using innovation management techniques appropriate for the companies R&D portfolios.

Innovation Management

Innovation Management combines the management of innovation processes as well as the management of change and development within the organization (Change Management) in the context of product-, process- and organizational innovation.

Innovation Management facilitates the ability of an organization to identify an opportunity, and to create and introduce new strategies, processes, or products.

One of the key challenges to Innovation Management within mature R&D organizations is the risk of complacency on the part of the R&D team, and a lack of willingness to change.

To reduce this risk, mature corporations can consider introducing a degree of entrepreneurial and innovation culture within their R&D teams, and start-up best principles can be used as inspiration.

While innovation sentiments often receive approval from top management (responsible for corporate vision), middle management (responsible for compliance with budget and productivity) is generally less willing to deviate from quantitative performance measures (such as Key Performance Indicators, KPIs).

For Innovation Management, some form of quantitative innovation measurement is thus needed to ensure that performance objectives are reached, while providing conditions conducive to leading innovation in R&D.

Innovation Management and the measure of the innovation performance challenges conventional Project Management techniques and organizational structures.

Addressing these challenges requires an innovation perspective, and successful Innovation Management ensures managers and R&D teams are able to cooperate with a common understanding of processes and goals.

• • •

Optimizing the matrix organization

Giving researchers the autonomy to introduce and test innovation through development methodologies based on iterative development (such as AGILE methodology) and pilot projects, and to create alliances and network across the organizational matrix can contribute to developing an innovative, entrepreneurial mindset within the R&D organization.

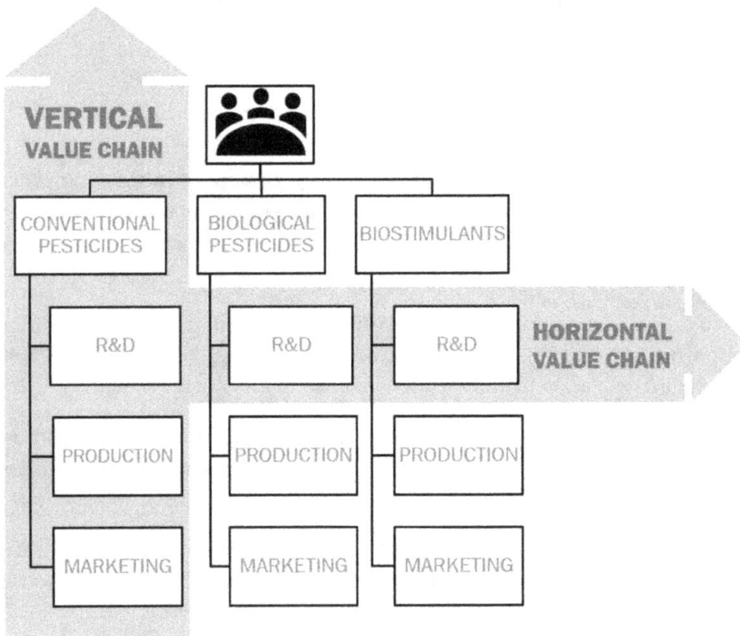

Figure 10.1: Vertical and Horizontal value chains in a matrix organisation.

Innovation alliances and networks may lead to the need to formalize cross-matrix alliances and introduce new R&D teams.

Within organizational matrixes, *vertical value chains* (Figure 10.1) ensure fast and targeted innovation for specific end-products, but tend to provide few synergistic opportunities.

In cross-matrix alliances, *horizontal value chains* may slow down product innovation, but may offer many synergistic opportunities.

For companies investing heavily in innovation and development, it may be necessary to radically restructure the organization of the company to adapt to the requirements for both incremental and iterative development frameworks.

In the 1990s, Lufthansa successfully implemented this strategy in a turnaround of the failing company, by "flipping" their organizational structure in a process of strategic renewal.

Individual functions, such as catering and maintenance (previously part of the *horizontal* value chain) became autonomous business units (for example, SkyChefs AG and Lufthansa AG) and thus components of a fast and targeted *vertical* value chain.

Innovation Management processes

For mature companies, incremental development of existing assets can appropriately be managed by so-called Gated Processes.

However, overly strict adherence to a Gated Process may lead to an inability to significantly develop the desired innovation outcome, resulting in the rejection of a project due to its inability to deliver a viable competitive product.

A Gated Process is part of an innovation or project management framework in which a development process is divided into distinct stages, separated by binary decision points (go - no go) known as "Gates".

For Gated Processes in the development of Defensive Core Portfolios, innovation is typically incremental (the linear optimization of the existing assets and strategies), and Time (efficiency) is an appropriate milestone of success.

These processes are typically driven by economic considerations and measured in terms of quantitative, output-based Key Performance Indicators (KPIs).

For these incremental innovation strategies based on a need to defend and develop a core business portfolio, strategic communication has traditionally been top-down in accordance with a "command and control" leadership mindset, coupled with a task-orientated skill set on the part of the "bottom" or executing organizational level.

In these organizations, business strategies are typically determined by economic Risk and NPV (Net Present Value) analyses, as positive NPV increases the value of the company.

However, NPV is inherently deficient as an *innovation* portfolio management tool. NPV analyses do not sufficiently take strategic position into account - for example, they typically do not account for option value ("can we use this in the future?"), synergy or cannibalization.

Other conventional management tools such as Scoring Matrixes (the analysis and scoring of multiple averaged factors) essentially have the same flaws as NPV analyses, with the added complication of being based on averages.

In an extreme example, teleportation would score high on all market and strategic parameters (but low on a single factor: feasibility!), leading to high *average* rating and the risk of being chosen as an economically attractive (but decidedly unfeasible) development project.

Recent shifts in technology opportunities and market demands have created a need for increasing levels of disruptive development, requiring the inclusion of *iterative* innovation strategies ("Fuzzy-Gate" or Iterative Processes).

For new companies and start-ups, an Iterative Process for innovation management can be particularly appropriate.

An Iterative Process (ideate - develop - adapt - iterate) is part of an innovation or project management framework in which a development process is repeated such that its experimental findings are applied to the results of its preceding processes, to obtain successively closer approximations to the solution of a problem.

For Iterative Processes, innovation is measured in terms of Learning (recurring cycles of timely feedback and adaptation). Iterative processes are driven by Vision and may be measured by qualitative, input-based key performance indexes (Figure 10.2).

	Iterative	Gated
Change	Disruptive	Incremental
Measure	Fuzzy (iterative)	Gated (binary)
Scoring	Feedback	Efficiency
Milestone	Learning	Time
Driver	Vision	Budget
Metrics	Qualitative KPI	Quantitative KPI
Leadership	Vision & delegation	Command & Control
Communication	Shared	Top down
Mindset	Entrepreneurial	Business
Skillset	Process orientated	Task orientated

Figure 10.2: Iterative vs. Gated innovation processes.

Iterative innovation strategies often require an entrepreneurial mindset across the organization, with shared strategic communication as a vital component to ensure integration of feedback into the business strategy.

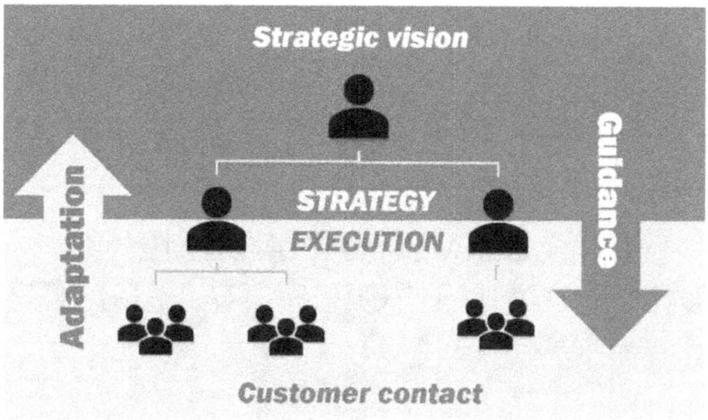

Figure 10.3: Ideal organizational structure for companies investing in innovation and development.

In these organizations, "vision and delegation" leadership mindsets are coupled with a process-orientated (rather than task-orientated) skill set on the part of the "bottom" or executing organizational level.

Accordingly, the organizational structure of a company often determines the type of innovation process possible.

In matrix organizations, top-level corporate management provides strategic vision and guidance to the organization by communicating business strategy, and by mapping business strategy to functional strategy.

Bottom-level experts or "executors" act as sensors with respect to client response and needs, and upward communication must be facilitated to ensure the integration of Learning (Adaptation) into business strategy, through the implementation of *Inclusive* Strategic R&D Management frameworks.

11

INCLUSIVE MANAGEMENT FRAMEWORKS FOR STRATEGIC R&D

A key issue in Crop Protection R&D is the mediation of biology-based strategies to conventional (chemical)-based decision-makers.

Crop protection biologists have a unique opportunity to mediate an understanding of the drivers of biologicals efficacy, and to ensure their implementation in the development of commercial strategies.

For many conventional companies, this can be achieved through the implementation of Inclusive Strategic R&D Management frameworks.

To ensure the inclusion of expert biological knowledge into R&D and business strategies, it is necessary to evaluate the companies' strategic framework and to consider implementing some degree of change management.

Traditionally, crop protection companies have favored incremental, product-orientated development.

Management levels have been responsible for defining the company's vision and strategy and for communicating this strategy in a "top-down" or "management-down" framework.

In this framework, execution of the strategy is carried out by the "down" or "expert" level. This level typically has the greatest customer or end-user contact, and receives important information necessary for product development.

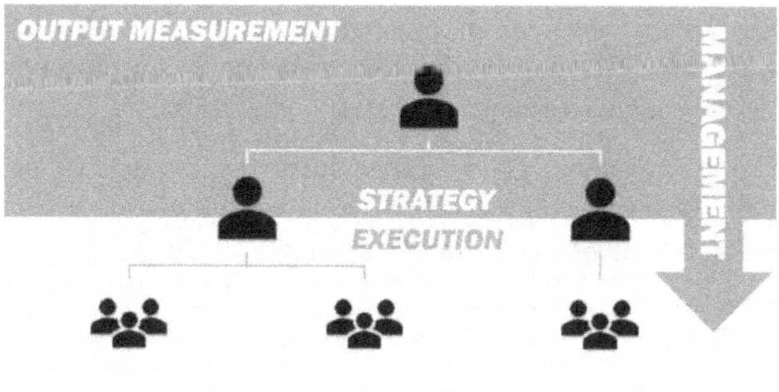

Figure 11.1: Top-down management framework.

Conventional output-based project management tools and evaluations (*e.g.*, Key Performance Indicators, KPI) are considered appropriate for incremental, product-orientated development frameworks, and are especially relevant for the "top" or management team (who after all are responsible for product output).

When companies disrupt their core business and introduce *innovation projects*, it may become necessary to revise this development framework.

An *inclusive (input-based), iterative R&D Management framework* and input-based evaluation can be introduced for innovation projects, to maximize the learning process while motivating the "down" or "executing" team.

An example of an iterative, input-based inclusive development framework is *Fair Process Management* (FPM) - a process developed by INSEAD (Blue Ocean Strategy) to facilitate inclusion of Execution (bottom-up) into Strategy (top-down) by creating individual buy-in at the beginning of the process.

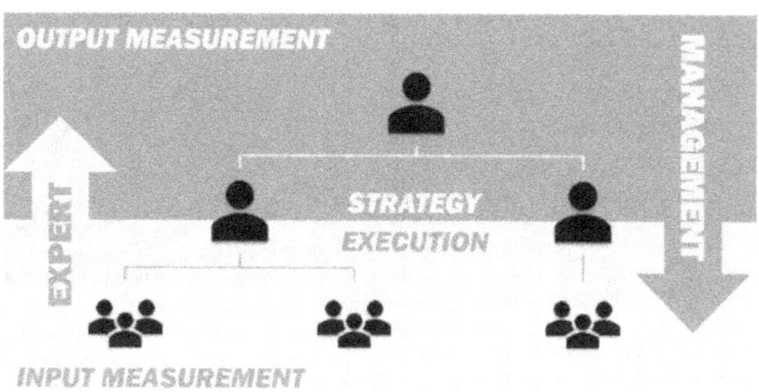

Figure 11.2: Inclusive management framework.

The objective of Fair Process Management is to optimize strategic decisions by management (the "up" team) as well as commitment from those involved in execution (the "down" team).

To facilitate the implementation of an input-based, iterative development framework (Figure 11.2), it is necessary to ensure open communication at the critical interface between strategy (vision) and execution.

For the development of biology based products, this requires the *"buy-in"* and *inclusion* of biology experts for strategic decisions, to the same level as chemists and chemical engineers who drive strategic decisions for conventional products.

"Great things in business are never done by one person. They're done by teams of people."

- STEVE JOBS

The objective is to provide an inclusive, iterative development framework to promote independent innovation initiatives and autonomous strategic biological input - "Frame Strong, Manage Light".

To achieve this, biological development groups should not function as support functions, but should be elevated to an increased strategic responsibility within conventional crop protection companies (Figure 11.3, left).

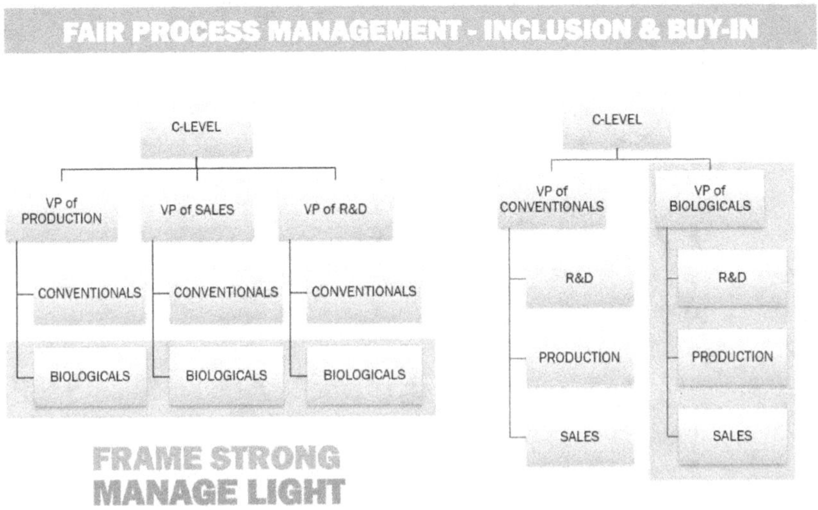

Figure 11.3: Organisational frameworks for the integration of biologicals into conventional crop protection organisations.

For companies investing heavily in biologicals development, it may be necessary to radically restructure the organization of the company (Figure 11.3, right) to adapt to the requirements for both incremental and iterative development frameworks.

A radical reorganization makes even more sense when we consider the economic and temporal differences between conventional and biological crop protection development.

For conventional chemistries, development times can be three times longer than for biological crop protection development, while expenses may be as much as 10 times greater.

Because of these differences, start-ups and other smaller, innovative companies have created a powerful and competitive environment for biopesticide R&D.

These small companies face a challenge in competing with incumbent agrochemical technologies, when attempting to persuade risk-averse growers to transition to this emerging technology, but also with respect to the economic requirements for efficacy testing, registration and distribution.

Current acquisitions, collaboration- and licensing agreements bear witness to the ongoing symbiosis in the crop protection market, and the need for Inclusive Strategic R&D Management frameworks as well as the mediation of biology-based strategies to decision-makers is greater than ever before.

12

FRONT LOADING AND VISUAL PLANNING IN STRATEGIC R&D MANAGEMENT

In today's competitive crop protection world, the rapidly changing strategic situation requires organizations to demonstrate strategic agility to transform strategies into action.

Within the R&D organization, introducing faster innovation processes with more predictable output is a strategic necessity. Put in simpler terms, "time is money" - this is especially true for innovation. We can illustrate the effect of the speed of innovation processes on business creation with an example:

In 2005, a leading agrochemical company launched an innovative herbicide strategy. However, because of poor communication between R&D and field application, the product did not address customer needs in the field. These issues required an extensive training period for the commercial team, leading to a missed season and delaying product introduction until the following year.

This delay allowed a competitor to exploit the situation by developing and launching a competing product addressing actual customer needs in time for the growing season. Because of their innovation agility, they could command a significant market share, and the company with the initial innovation experienced an estimated 5 million Euro in lost revenue.

Late-stage changes are costly and delay product development

In the crop protection industry, almost everyone involved in product development has experienced late-stage changes to address prototype failures, patent challenges, manufacturing issues, production up-scaling or efficacy issues in the field.

For most development projects, the cost to make a change during the initial phase is lower than during the prototype phase, which in turn is significantly lower than in the production phase.

Despite the cost of changes increasing over product development project timelines, for many crop protection organizations, late-stage changes are still considered inevitable and are factored for in project plans.

For crop protection organizations, reducing the number and impact of late-stage changes can be achieved by "Front Loading" the product development process - moving critical knowledge to the beginning of the project.

Front Loading takes place at the initiation of the project when the ability to carry out changes is high and the cost of these changes is low, and when decisions have the most influence.

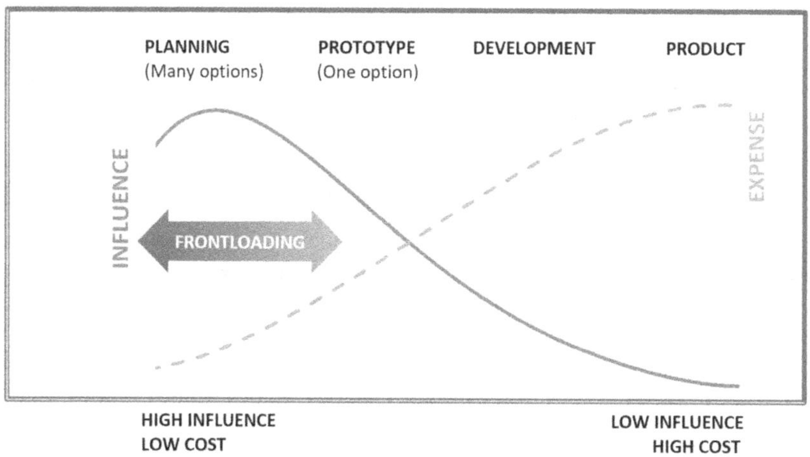

Figure 12.1: Influence and expense profile - Front Loading reduces late-stage changes by moving critical knowledge to the beginning of the product development process when decisions have the most influence with the least expense.

Front Loading facilitates the definition of objectives and associated planning processes early in the project life cycle. As we make the most crucial decisions at the start of the project, Front Loading critical knowledge to the beginning of the project will reduce the number and impact of late-stage changes.

While Front Loading adds time and some expense to the first stages of a project, these costs are relatively minor compared to the costs and consequences of changes made at a later stage.

Doing the right thing - targeting customer value

R&D Front Loading is a crucial part of Lean Innovation - which may be defined as both 'doing the right thing' and 'doing it right':

Doing the right thing means ensuring that the outcome of the project targets the customer need

Doing it right entails implementing Front Loading decisions through a prompt and efficient development process

We can summarize this as: "Creating more value for the customer, with fewer resources."

In Lean Innovation, the innovation project team identifies the solution generating the greatest customer value by identifying and targeting customer needs, while avoiding over- or under-developing the product.

In successful R&D organizations, management and project teams have a well-defined "top-down" and "bottom-up" process for innovation projects.

This input-based, inclusive development framework facilitates the inclusion of Execution (bottom- or expert-up) into Strategy (top- or management-down) by ensuring individual buy-in at the beginning of the process.

The textbook example of the value of identifying customer need in Lean Innovation is that of the redesign of the Toyota Sienna minivan for its primary market - the United States.

Rather than relying on surveys, Toyota addressed customer needs by sending an engineer to the US to rent current model Siennas and drive 85,000 km throughout North America, from Canada to Mexico.

This focus on identifying customer need meant crosswind stability and steering drift were improved for long-distance travel, while the turning radius was tightened for city driving.

In addition, the number of cup holders was increased to 14(!) and the fold-down load area was increased to 4 by 8 feet to transport plywood and drywall sheets. The result: Sienna sales increased by 60 percent

As a Strategic R&D Manager, you have responsibility for the innovation process. However tempting it is to push the project team to start the value creation process, it is crucial to first define the customer need to be targeted by the innovation process.

It is critical for the process that you define, clarify and align internal and external expertise with project objectives and success criteria by:

Establishing customer value to cut "solution waste" and "solution creep" - **the outside-in approach to innovation**

Empowering cross-functional teams to provide critical knowledge up-front and challenge conceptions of customer value - **the inside-out approach to innovation**

The key benefit of cross-functional teams in knowledge-based development projects lies in implementing cross-discipline knowledge transfer from earlier projects and transferring problem and solution-specific information to new projects - transforming knowledge into value.

For example, process engineers can inform the team on the process capability of existing manufacturing equipment at the start of the project.

Trial specialists can provide information on testing methods and time constraints to ensure timely production of prototypes for field trials, while agronomists can give information on competitor products and regional requirements and limitations.

The project team can plan within these constraints to find and solve development problems before they arise, improving product quality and project efficiency.

Implementing knowledge-based projects requires "translating" knowledge from specialist terms into general terms that the team can understand, increasing cross-functional expertise within the organization so that this knowledge is ready to be implemented in later product development projects.

Let us consider two project teams tasked with the same goal of developing an innovative crop protection solution.

Team A has a clear, well-defined commercial goal, the result of an intensive Front Loading process. They are a balanced cross-functional team with strong individual competencies and the correct combination of skills as a team, combing patent, biology, registration, synthesis, formulation, production, commercial and marketing specialists. Together with empowerment, this combination of clear objectives and competencies makes for a motivated team willing to challenge management vision when needed. By using Front Loading templates, team members provide input before project meetings, resulting in structured project meetings that give value to all participants. Project changes and iterations are made early in the project process, and the project milestones, budgets and deadlines are met.

Team B starts with a less well-defined goal. Despite their individual competencies, the lack of Front Loading objectives has led to the prioritization of department deliverables, rather than project deliverables. Due to the low level of up-front knowledge and unstructured project meetings, project changes and iterations are made later in the project, and project milestones, budgets and deadlines are exceeded.

Team A's development process is clearly better. The motivational project culture and associated processes make sure that relevant knowledge is integrated into a solution creating customer value by balancing capabilities with market objectives.

In a knowledge-based crop protection organization, the management team is responsible not only for the strategic vision but also for establishing the culture and associated processes that drive commercial innovation projects.

Doing it right - Front Loading Crop Protection R&D

One challenge of Front Loading is deciding what and how much information we need. After defining the project (the "What" and the "Why"), the joint project team (representing, e.g., patent, biology, registration, synthesis, formulation, production, commercial and marketing) will brainstorm questions such as customer value, markets, patents, regulatory requirements, production, NPV (Net Present Value) and COGS (cost of goods sold).

Some questions might be answered at once; others might need to be answered through estimates. For verification of estimates, and to answer the remaining questions, team members will disperse to obtain the remaining answers, to be shared with the rest of the team at a later meeting.

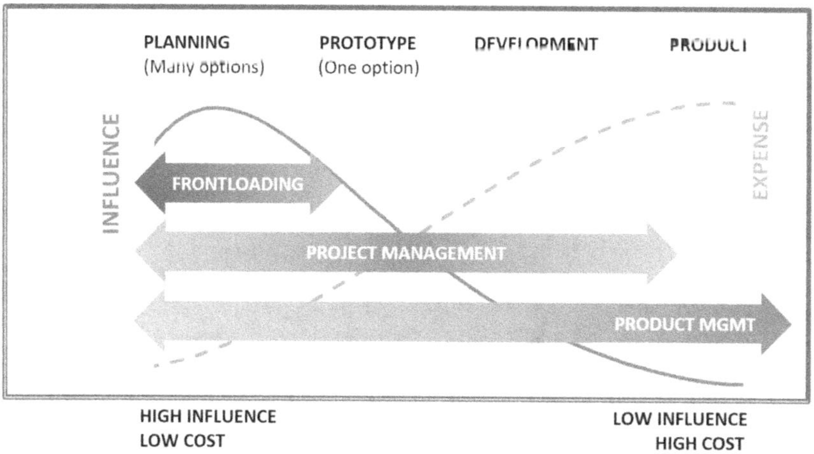

Figure 12.2: Cross-functional Front Loading introduces critical knowledge at the beginning of the product development process, when decisions have the most influence with the least expense.

All team members need to hear questions from other departments - not just to facilitate agreement, but also to reduce the risk of late-stage changes.

In one example, at an early Front Loading meeting the commercial specialist asked what the requirements for a Brazilian product launch would be.

This caused some surprise in the project team, as the Brazilian market was not included in the project description and had not otherwise been mentioned. This example shows the importance of Front Loading with participants who would otherwise be involved in the project much later.

Front Loading gives the team the chance to discuss critical changes at the start of the project and avoiding expensive delays at the end.

This process becomes easier as the organization gains experience in Front Loading, and a template to facilitate future Front Loading can be introduced.

Using a template containing the most relevant questions will allow team members to document critical knowledge before the first Front Loading meeting, making the decision process more efficient.

The following table contains examples of typical Front Loading questions for crop protection development projects:

Objectives	Do your strategic objectives address an actual need (do we have buy-in from regional commercial managers? What are your competitors doing that gives them a strategic advantage? What is the value of this strategic objective - how will the outcome justify the cost of investment (has an NPV been prepared)? What are we delivering - what are we not delivering? How will a successful outcome be defined - what is the benchmark for success?
Stakeholders	Who will decide? Key stakeholders - patent, biology, registration, synthesis, formulation, production, commercial and marketing - who else should we involve?
Timeline	What is the Timeline - what are the Key Milestones? Deadline? GANTT diagram
Challenges	What problems do we face? What are the top issues that are acting as barriers to success – what might get in the way (staffing, timeline, budget, technical issues, patent, registration)? What do we need to strengthen to support achieving this strategic objective?
Resources	Detailed budget per cost center (from NPV) Hours (est.)

The next table contains examples of Front Loading questions for biological evaluations.Here, Front Loading will typically take place between the requisitioner (for example, Formulation Development) and the biological testing facility (e.g., the greenhouse team).

As a manager of both cross-functional innovation project teams and biological evaluation teams, I have experienced that Front Loading increased the overall rate at which we identified and solved development problems.

This was in part because we had all the critical knowledge up front, allowing us to estimate resources and timelines effectively and saving a lot of wasted trials.

This information had always available within the organization, but by communicating systematically and effectively "up-front," we experienced a significant decrease in waste, and an increase in project effectiveness and success.

Following the Front Loading process, we are ready to move to the second phase of LEAN innovation: implementing Front Loading decisions through a quick and efficient development process.

Objectives	What is the aim of these trials? Is the aim SMART (Specific, Measurable, Achievable, Relevant, Time/cost-bound), e.g., improve rainfastness from 4 hours to 2 hours after application What are we delivering - what are we not delivering? How do we define a successful outcome - what is the benchmark for success?
Challenges	What problems do we face (e.g., staffing, timeline, budget, technical issues, other)?
Timeline	Study initiation? Trial start? Trial completion? Study completion?
Resources	Cost center Hours (est.)
Active ingredient	Mode of action; logP; pKa; solubility, etc. Safety data sheet available?
Formulation	Type; concentration? When will first formulations be ready? How many formulations are we to test? Reference products? Who will supply references? Safety data sheet available? Additional instructions regarding suspensibility, stability, pH requirements, adjuvants, etc.
Trial	Define type, e.g., Soil degradation - eye toxicity - crop safety-residual activity - efficacy - persistence - rainfastness - other?
Target species	Pests, weeds, or pathogens. Specify, e.g., sucking insects (aphids) and grazing insects (mites). Additional information: e.g., mites feed on the underside of leaves - ensure full leaf coverage at spraying
Target crops	Broadleaf or grass crops; pre- or post-crop-emergence? Specify, e.g., post-crop-emergence in cereals (grass crop). Crop safety evaluations needed? Additional information: e.g., barley is more sensitive to phytotoxicity than wheat.
Application (ranked in order of relevance)	Soil - Foliar - Seed treatment - Spray application? Application volume (e.g. L/Ha) Rate (e.g. g ai/Ha) Timing (e.g. pre-, peri-, post-emergence; protective or curative)
Evaluation	Type and timing, e.g., Visual evaluation, % PHYGEN at 7, 14 & 21DAT (EPO)
Data requirements	e.g., randomized pot trial; 5 replicates Analysis: One-way analysis of variance (ANOVA); Tukey post-ANOVA comparison-of-means.

Do it right - Visually Planning Crop Protection R&D

Visual Planning is a resource management platform that facilitates the implementation of Front Loading decisions through a quick and efficient development process.

In the following example of biological testing project management in a laboratory or greenhouse facility, we use a Visual Planning board to manage incoming tests within the lab's daily operations schedule.

The planning board provides a visual overview of lab capacity as well as an outline of work in progress and work to be done.

Visual Planning boards can be physical (e.g., Post-Its or magnets on a whiteboard) or virtual (Excel spreadsheet).

The advantage of a physical board is that it is always visible for all team members, while the disadvantage is that updating the board is a manual process that takes time.

The virtual board has the advantage that many processes can be automated, while the downside is that it is not always visible.

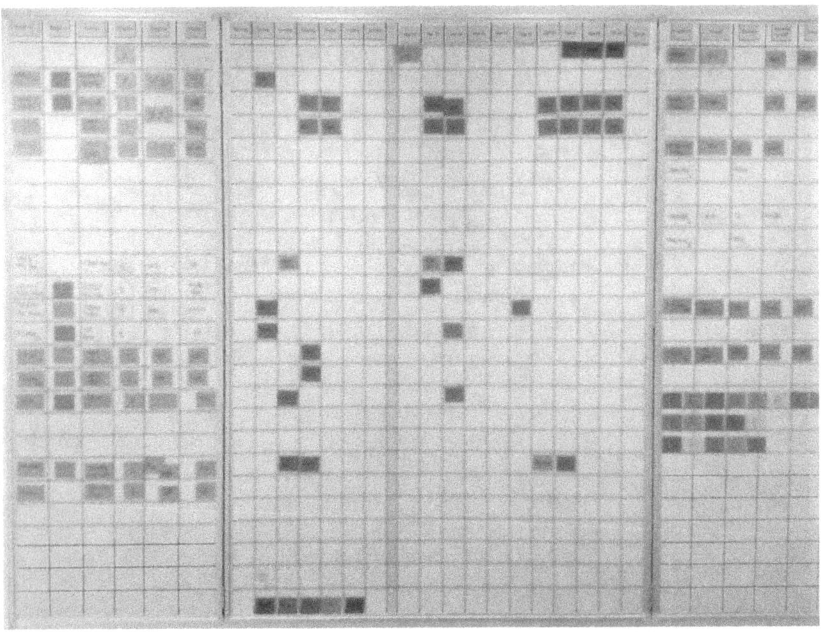

Figure 12.3: Physical (wall-mounted) Visual Planning board using a whiteboard and magnets.

Thus, the virtual board is suited for individuals or small teams whose activities are computer-based, while the physical board is suitable for larger groups in a more hands-on lab or greenhouse environment.

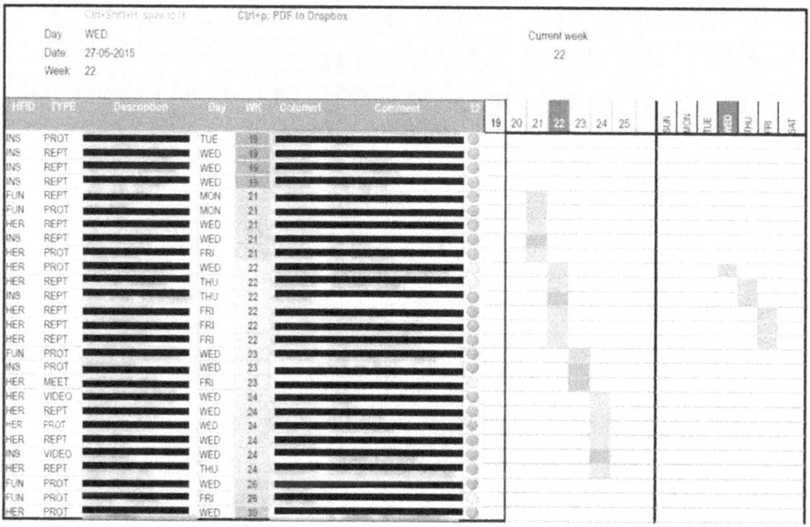

Figure 12.4: Virtual (spreadsheet-based) Visual Planning board

A further advantage of the virtual board is that completed tasks can be archived and used to prepare summaries of annual task distribution, completion statistics, resource use, and other key performance indicators.

For lab and greenhouse testing facilities, we can divide the board into four sections:

Work in progress Task section: project ID; aim; priority; deadline; requisitioner; internal responsible

Work in progress Daily planner: Monday to Sunday

Work in progress Weekly planner: 8-12 weeks using ISO week-numbering

Work to be done: project id; aim; expected start data; deadline; requisitioner

---

The x-axis of Sections 2 (day) and 3 (week) cover the time dimension, with one box for each day or week, while each y-axis row covers one task trial.

We can obtain an overview of task distribution (Section 1) using colored magnets - for example: white for insecticides, silver for fungicides, and gold for herbicides. In the daily and weekly task sections, colored magnets for each member of the team provide an accessible resource distribution overview.

New requisitions for trials are placed in Section 4 (work to be done). At the weekly planning meeting (30 minutes), these trials are designated to appropriate weeks, depending on resource availability and task priority, and an internal responsible team member assigned.

By preparing and maintaining the Visual Planning board, lab staff accept ownership of the plan, understand the overall strategy, and take responsibility for achieving milestones on time.

Ownership of the plan relieves the project manager from "dictating" deliverables, and lab staff become accountable to external members of the project team tasked with providing deliverables on time.

By involving external project team members, lab staff are aware of the cross-functional needs of the project. Other advantages include:

Team members feel ownership for the individual trials and have a stronger sense of responsibility

Management has an overview of work in progress, as well as any issues

Management becomes visible and directly involved in daily operations

External project team members can evaluate resource availability and follow the progress of their prototype

---

In this environment, the project manager can focus on managing the overall process, rather than the individual steps. This includes providing perspective, steering discussions and maintaining focus, promoting creativity, and securing additional resources when needed.

When implemented successfully, the Visual Planning process becomes the lab team's responsibility. It is the lab team who should decide the timing and agenda of the regular weekly meetings.

In this way, the lab team can adapt the Visual Planning board to their requirements, increasing their overview, reducing lead times and backlogs, and improved scheduling and resource planning.

In summary, Front Loading ensures business-driven objectives and associated planning processes are introduced early in the project life cycle, when the ability to carry out changes is high, and the cost of these changes is low.

Visual Planning is a resource management platform that facilitates the implementation of Front Loading decisions through a quick and efficient development process. Together, Front Loading and Visual Planning ensure faster, effective innovation with more predictable output.

13

CASE STUDY: STRATEGIC R&D MANAGEMENT

The following case study is fictitious and based on experiences and observations made across the crop protection industry, and does not represent the experiences of any specific company.

The case study begins with the initial strategy for the development of a new biological product by AGCHEMCO, a conventional (chemical) crop protection company. It presents and analyses the problems encountered through the implementation of a conventional development strategy, and describes how this learning process was applied to develop a new biological development strategy.

In December 2013, the CEO of AGCHEMCO hosted a meeting of the company's senior management. The order of business was simple: future sales estimates for the companies agrochemical products were in decline, and the immediate implementation of a revised business strategy was imperative.

Stricter pesticide regulatory requirements in the EU and Brazil had restricted market access and increased registration costs, while an associated decline in the number of registered active substances in the EU was reducing the viability of crop protection resistance strategies.

In addition to stricter pesticide regulatory requirements, consumer awareness regarding hazards caused by chemical pesticides, and the phasing-out of toxic conventional pesticides contributed to the decline in sales estimates - these same factors were driving the growth of the agricultural biopesticide market.

A drastic strategy was presented: the immediate implementation of drastic cost-cutting measures and the culling of low-profit product lines to create a leaner, more effective organization.

In addition to optimizing its core business, AGCHEMCO would enter the rapidly growing biopesticide market by leveraging its existing agrochemical market access and development experience.

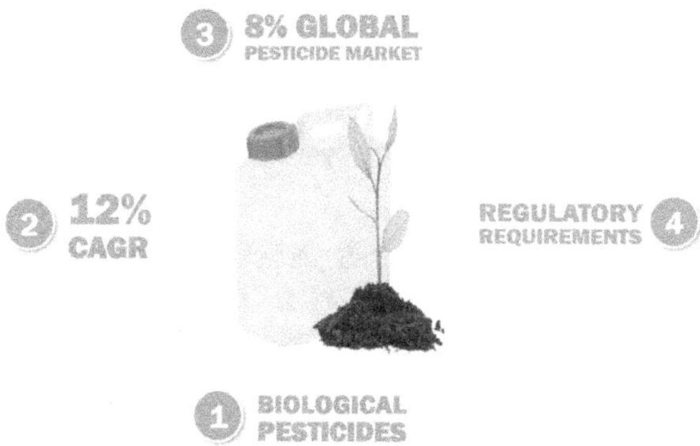

Figure 13.1: AGCHEMCO's case for entering the Biopesticide market.

Negotiations to access a promising first-stage biological fungicide candidate had been successfully completed, with the objective of completing in-house development and registration and attaining product-to-market within three years.

The strategic reasoning was sound. Estimates of global market sizes for crop protection products (conventional pesticides as well as biopesticides) varied, but the general consensus was that biopesticides were expected to increase from about 2% of the global pesticide market in 2003 to about 8% of the global pesticide market (estimated to exceed 82 billion USD) in 2020.

The prevailing opinion was that the compound annual growth rate (CAGR) for biopesticides during this period was expected to exceed 12%, compared to an expected overall pesticide CAGR of about 7%.

The proposed vision and strategy was received positively by Executive Management, and the VP of R&D was assigned the responsibility for its execution.

Her first task was to incorporate a Biologicals project group into the existing

R&D structure for the companies conventional crop protection organizational framework:

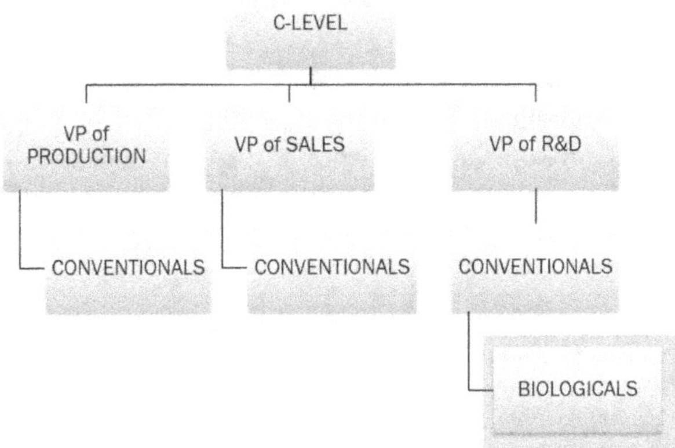

Figure 13.2: Incorporation of the Biologicals Project group.

The project group was headed by a formulation chemist and included a production chemical engineer as well as an agrobiologist.

R&D Management convened a meeting of the project group and introduced the new biofungicide candidate - a microbial fermentate produced in pilot batches. The acquired data package was sparse, and the only available efficacy data was from initial lab trials.

The objective of the project group was defined: the development of a viable, competitive product - successfully tested for efficacy under lab and field conditions. The final product would be registered and in the market in three years, leaving only twelve months for development.

Standard project performance parameters including deadlines and KPIs were defined to ensure that outputs could be reached on time.

To be successful, the biofungicide would need to compete in the cereal fungicide market, competing against conventional (chemical) products with both protective and curative activity.

Due to the perceived variability of biopesticide efficacy, the biofungicide candidate would need to work at least as well and as consistently in the field as the commercial references.

Over the following four months the Formulation group developed a stable formulation of the biofungicide, for which efficacy was confirmed in a series of initial agar-plate and leaf-disk assays.

Following the initial round of testing, whole-plant greenhouse and field trials were carried out. After the third set of greenhouse trials, R&D Management met with the project group.

Many results were disappointing. The initial experiments had confirmed and replicated the original lab data, and had demonstrated that the formulated product was more stable than the original microbial fermentate.

The initial trials were scaled up to whole plants in the greenhouse, and comprised protective applications, with treatment 24 hours before pathogen inoculation, as well as curative applications with inoculation 24 hours after treatment.

The commercial references had worked well, but for the biofungicide formulation fungicidal activity was only apparent at such high application rates that the test plants suffered significant leaf damage.

Data from field trials proved equally disheartening. While some of the trials had been a success, an equal number showed no effect at all.

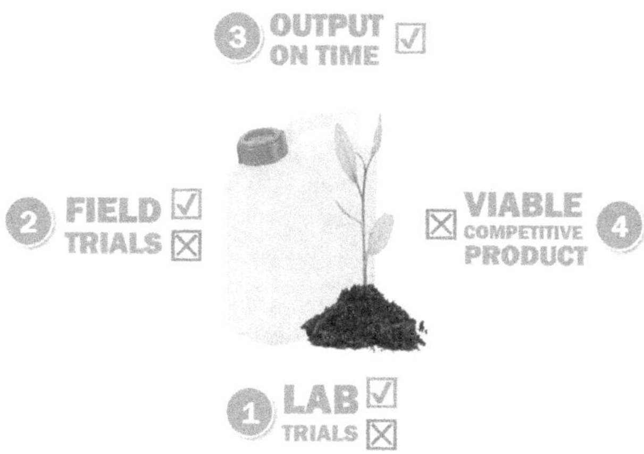

Figure 13.3: R&D status for AGCHEMCO's initial Biofungicide development testing.

Although the formulation had been shown to be stable, and project milestones on the formulation work and efficacy testing were met - the efficacy data was so variable that the goal of a viable competitive product had clearly not been reached.

At the October 2014 AGCHEMCO Executive meeting, results of the core chemical crop-protection business optimization measures were presented.

The implementation of drastic cost-cutting, the culling of low-profit portfolios and the inception of strategic alliances promised a leaner, more effective organization.

For the first time in years, the commercial estimates for the companies core agrochemical products were no longer negative.

Following the presentation of the core business results, the status of the biologicals development project was the next order of business.

A structured analysis of Strengths, Weaknesses, Opportunities and Threats (SWOT) to identify the situational viability of introducing a biological product to AGCHEMCO's portfolio identified the weakness that gave greatest cause for concern: the variable efficacy of biological crop protection products under field conditions.

With the decline in their core business halted, AGCHEMCO debated dropping their foray into biological crop protection in order to focus resources on the optimization of their conventional agrochemical portfolio.

To make this decision it would be necessary to analyze what had been learnt, and to consider whether a revised biological development strategy could lead to the successful development of a biological portfolio.

Lessons Learnt

An initial finding was that existing biological knowledge had not sufficiently been incorporated into the product development strategy. This led to inappropriately designed lab and field efficacy studies.

By focusing on competitiveness against conventional products and using standard protocols for conventional chemical products, trials had been set up as protective and curative essays, with application timings 24 hours before inoculation for protective assays, and 24 hours after inoculation for curative assays.

This timing did not allow for the extended kinetics of plant defense introduction necessary for optimal biopesticide efficacy, leading to poor efficacy compared to the commercial chemical reference products.

Subsequently, it was understood that the kinetics of plant defense responses lead to the biopesticide performing adequately in petri-dish but not in whole-plant essays.

Furthermore, this provided an explanation for the variability of field trials:

trials in regions exposed to disease pressure immediately after the application of the product showed low levels of efficacy (there was no time for the induction of plant defense responses), while trials carried out in regions where inoculum appeared later showed greater levels of efficacy.

There was initially little incentive or opportunity to challenge the development process by integrating biological learning into the development strategy, as the performance of the project was measured on output based Key Performance Indicators, KPIs.

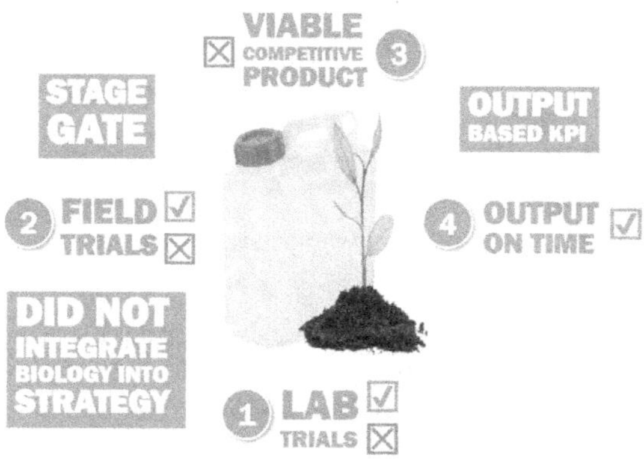

Figure 13.4: Issues in the development of a new biological product using an incremental R&D framework.

In addition, the stage gate nature of the project did not provide resources or flexibility for iterative development i.e. the implementation of experimental learning into the redesign of new trials.

In strictly adhering to a stage gate process, the inability to significantly develop the product could lead to the rejection of the project based on its inability to deliver a viable competitive product.

For innovation projects, an iterative gate process can be more appropriate. In our example, iteration and the integration of biology into strategy would focus on optimization of the application of the product, rather than optimization of the product itself.

To understand the value of integrating biological ("bottom-up") input into the strategic development process, we should consider a SWOT analysis (identifying factors favorable and unfavorable to the development and application of biopesticides), and to interpret it from the point of view of a biologist, as opposed to the conventional point of view of a chemical engineer.

STRENGTHS	WEAKNESSES
• Zero-day pre-harvest interval • Use under moderate to severe disease pressure • Tank mix / rotational program candidates • Quicker registration / organic farming • Can be cheaper than chemical pesticides • Can be more effective than chemical pesticides • Biodegradable; less likely to harm non-target species	• High specificity • Often slow speed of action • Often variable efficacy • Target resistance • Issues with lifespan & application technology
OPPORTUNITIES	**THREATS**
• Seed treatments and soil amendments. • Biostimulation against biotic & abiotic stresses • IPM; Mass production; fermentation • Formulation and application technology • New technology: growth potential	• Grower objection • Regulatory changes • Plant-defence toxins (current debate on toxins in organics) • Counterfeit CPP

Figure 13.5: SWOT matrix for the inclusion of biopesticides in crop protection strategies.

From a biological / end-user point of view an important strategic strength of biopesticides lies in their application at very specific timings: short pre-harvest intervals allow the application of biopesticides (especially biofungicides) immediately before harvest, where very few (if any) conventional pesticides are available.

Biopesticides (especially bioinsecticides) known not to harm non-target (beneficial) insect species may be applied during flowering, a period where many conventional pesticides are not able to be used due to their effect on bees.

Accordingly, if the biopesticide candidates are to be strategically placed to utilize these windows of opportunity, there is not much point in comparing their efficacy with commercial references which are not relevant.

Biostimulation and protection against biotic and abiotic stresses have been reported for many agricultural biologicals. Inclusion of biostimulation and stress alleviation evaluations in the design of biopesticide efficiency trials provides important decision-making parameters for the strategic viability of the tested product.

Often, the only biological parameter considered when evaluating the product is its speed of action and efficacy. However, an understanding of the kinetics of plant defense induction will allow the design of appropriate trials, and the inclusion of appropriate measures of comparison.

Accordingly, for this case, curative experiments might have been eliminated, as many biopesticides attain much of their effect through the induction of plant defense response.

As these responses must be induced in advance of the pathogen reaching the leaf surface, protective applications are considered most appropriate for these products.

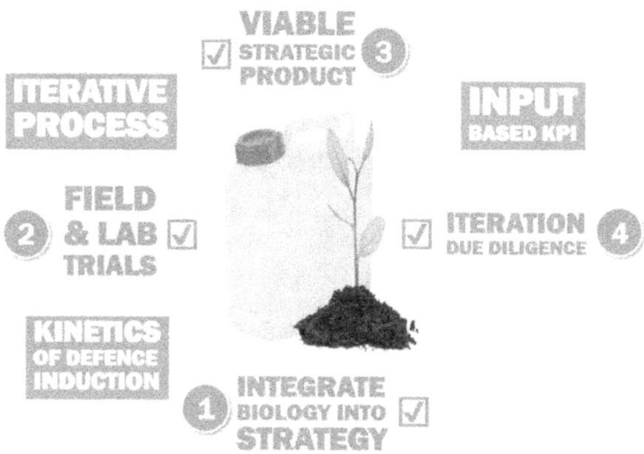

Figure 13.6: Learnings - the development of a new biological product using an iterative R&D framework.

Rather than performing curative trials, resources could be allocated to optimizing the induction of plant defense responses, for example by formulating the biofungicide in yeast cells (natural microcapsules), which contribute to the induction of defense responses (the yeast cells are recognized as possible pathogens) while protecting the biopesticide from desiccation and UV degradation.

———

PLEASE LEAVE A REVIEW

Thanks for reading THE LABCOAT GUIDE TO CROP PROTECTION R&D! Your support makes it possible for this author to continue creating.

If what you read provided value, **please leave an honest review** wherever you bought this book. Your feedback is invaluable, and reviews help new readers discover my work.

To get in touch with me directly, please use one of the various contact methods listed at http://biocomm.eu

www.ingramcontent.com/pod-product-compliance
Lightning Source LLC
LaVergne TN
LVHW052022200926
848481LV00005B/1563